DEC 30 1993

TOBY NADEL — ARCHITECT AIA
4304 EAST GENESEE ST.   DEWITT, N.Y.
(315) 446-2341   13214

DEC 30 1988

TOBY NADEL — ARCHITECT AIA

STP 1088

Roofing Research and Standards Development: 2nd Volume

Thomas J. Wallace and Walter J. Rossiter, Jr., editors

ASTM
1916 Race Street
Philadelphia, Pa 19103

Library of Congress Catalogue Card Number: 87–31938
ASTM Publication Code Number (PCN): 04-010880-57
ISBN: 0-8031-1393-5
ISSN: 1050-8104

Copyright © 1990 by the American Society for Testing and Materials. All rights reserved.
No part of this publication may be reproduced, stored in a retrieval system, or transmitted,
in any form or by any means, electronic, mechanical, photocopy, recording, or otherwise,
without the prior written permission of the publisher.

NOTE

The Society is not responsible, as a body,
for the statements and opinions
advanced in this publication.

Peer Review Policy

Each paper published in this volume was evaluated by three peer reviewers. The authors
addressed all of the reviewers' comments to the satisfaction of both the technical editor(s)
and the ASTM Committee on Publications.

The quality of the papers in this publication reflects not only the obvious efforts of the
authors and the technical editor(s), but also the work of these peer reviewers. The ASTM
Committee on Publications acknowledges with appreciation their dedication and contribu-
tion of time and effort on behalf of ASTM.

Printed in Chelsea, MI
November 1990

Foreword

The papers in this publication, *Roofing Research and Standards Development: 2nd Volume,* were presented at a symposium held in San Francisco, California, 17 June 1990. The symposium was sponsored by ASTM Committee D08 on Roofing, Waterproofing, and Bituminous Materials. Thomas J. Wallace, U.S. Navy, and Walter J. Rossiter, Jr., National Institute of Standards and Technology, served as symposium cochairmen and are editors of this publication.

Contents

BITUMINOUS ROOFING SYSTEMS

INDEXES

OVERVIEW

The revolution that has occurred in the U.S. roofing industry over the last decade is well known to all. Many roof systems which incorporate elastomeric, thermoplastic, and polymer-modified membranes were unheard of in the early 1970s, but are commonly used today. Improvements in their chemistry and production technology have enhanced the performance of these materials and made them economically competitive. Further, the use of glass- and polyester-based fabrics has essentially replaced the traditional reinforcements for built-up roofing systems. With the arrival of the new membrane systems, innovative methods of membrane securement such as ballasting and mechanical attachment were brought into the industry. A major result of the changes in roofing practices was increasing pressure on ASTM Committee D 08 on Roofing, Waterproofing and Bituminous Materials to develop standards to assist in the proper selection, use, and application of these systems. Committee D 08 has responded well. Many task groups have been formed and are diligently working toward development on the needed standards.

The Proceedings of two symposia describing the changes that have occurred and the needs for research to support development of standards were published in the 1980s -- Single-Ply Roofing Technology, ASTM STP 790 (1982), edited by W.H. Gumpertz, and Roofing Research and Standards Development, ASTM STP 959 (1987), edited by R.A. Critchell. But the work of Committee D 08 is far from finished, particularly in the area of conducting research to support the standards development process.

The members of D 08 firmly believe in the importance of having a strong technical basis for their Committee's standards. The availability of data can help accelerate the standards development process, since decisions can be made on fact and not opinion. In 1987, Subcommittee D 08.21 on Research Needs for Roofing and Waterproofing was formed to: (1) provide information on ongoing research, (2) identify research needed in standards development, and (3) disseminate research results through activities such as workshops and symposia. This symposium represents the first fruits of the Subcommittee's efforts to foster research in the support of standards. It illustrates the commitment made by D 08 to the improvement of roofing technology.

As in the past, this publication is dedicated to the members of ASTM Committee D 08 who give unselfishly of their time and energy to improve the performance of

1

roofs. The editors express their sincere thanks and appreciation to those many individuals who participated in the organization and conduct of the symposium. R.A. Alumbaugh, C.G. Cash, E.F. Humm, D.F. Jennings, C.F. Mullen, W.T. Rubel, T.L. Smith, and S.W. Warshaw were members of the Steering Committee. D.E. Richards was a session chairperson. Dorothy Savini, ASTM, provided for the symposium arrangements. Barbara Stafford, Therese Pravitz, Kathy Greene, and other ASTM staff members directed the review and publication of the papers. Finally, special thanks are given to the authors and reviewers of the papers without whose efforts the symposium would not have been possible.

Thomas J. Wallace
Naval Facilities Enginnering Command
Philadelphia, PA 19112
Symposium Co-Chairman and Editor.

Walter J. Rossiter, Jr.
National Institute of Standards and Technology
Gaithersburg, MD 20899
Symposium Co-Chairman and Editor.

Elastomeric and Thermoplastic Roofing Systems

Christopher P. Hodges[1]

Characterization of Lap Seam Strength for In-place and Laboratory Prepared EPDM Roof Membranes

REFERENCE: Hodges, Christopher P.,"Characterization of Lap Seam Strength for In-place and Laboratory Prepared EPDM Roof Membranes," Roofing Research and Standards Development: 2nd Volume, ASTM STP 1088, Thomas J. Wallace and Walter J. Rossiter, Eds., American Society for Testing and Materials, Philadelphia, 1990.

ABSTRACT: A series of T-Peel lap seam adhesion tests were conducted on various EPDM roof membrane samples in order to characterize the typical expected seam strength in laboratory prepared samples, and in samples taken from new and in-service roofs. The program included constructing lap seams in the laboratory using four commercially available EPDM roof membranes and their proprietary adhesives and seaming techniques. In addition, two sets of seams were made and tested in the laboratory without the benefit of cleaning the splice area of the rubber. T-peel testing was also conducted on samples taken from two roofs that were approximately 6 months and 5 years in age. Three other sets of contractor prepared field seams were obtained during on-site roof construction and tested in T-peel. The objective of the study was to provide some reference data relative to seam strength developed in the laboratory versus that which is typically achieved in the field.

KEY WORDS: EPDM, T-peel test, lap seam, adhesion, strength development, single-ply, field seams, butyl adhesive, neoprene adhesive, in-service.

In the roofing industry, the last 25 years have brought dramatic changes in the use of new materials and assemblies. A market that was once dominated by built-up roofing is now inundated with new products. One material, first introduced in the 1960's, has dominated the single-ply market for a number of years. EPDM (ethylene propylene diene terpolymer) accounts for over half of the single ply market.

[1] Senior Materials Engineer, Law Engineering, Chantilly, Virginia 22021

5

It has generally been accepted through a number of marketing and research efforts that, for single-ply roofs, the field seam is the most frequently reported problem area [1]. The primary method that has been developed to characterize the seam strength is the T-peel test (ASTM D 1876). Research has shown the T-peel test to be sensitive to typical application parameters [2,3,4].

Although the seam has been identified as the leading problem in single-ply roofing, we still do not fully understand and do not know the strength required of a single-ply membrane field seam under actual service conditions. A significant amount of data is available regarding single-ply membrane material properties, however, there is an overall lack of data about the minimum levels of mechanical properties required of a roof membrane in-service.

Through other research, and the efforts of this study, we begin to understand the typical expected "as-built" strength of an EPDM lap in T-peel. It would be inappropriate to assume that the failure mechanism of all in-service seams would necessarily be in peel. Due to imposed loads, a roof membrane may be subjected to combinations of shear and peel forces. Until such time as a test is developed which more accurately duplicates the typical failure mechanism, the T-peel test remains one of the most sensitive test methods to parameters such as seam cleanliness, application temperature, application rate (adhesive thickness), seam open time and other factors. In contrast to other construction materials, which typically measure their strength with significantly high numbers, the peel strength of an EPDM seam is relatively low, typically measured to be less than 1.75 kN/m (less than 10 pounds per inch of seam).

This study presents results of over 300 T-peel tests conducted on commercially available EPDM membranes and their associated adhesive products. The focus of the study was to provide further data on the characterization of seam strength values under laboratory and field conditions. The laboratory portion of the study was conducted on roofing materials that were purchased from local roofing suppliers, or were supplied by roofing contractors during a roof construction project. The purpose of this was to use actual "off the shelf" products to gain practical knowledge about how these products perform. The study was divided into three areas:

1. Analysis of properly prepared laboratory seams to characterize the development of strength at early ages, and establish some expected strength values.

2. Analysis of laboratory prepared seams which were made without using the recommended preparation and cleaning process. The purpose of this was an attempt at duplicating what might occur in the field if the applicator were to omit one of the recommended steps in the seam cleaning process.

3. Analysis of some lap seams that were obtained from roofing projects. This included testing of contractor - prepared lap seams taken from roofs during the

construction process, and testing lap seams that had been in-service from a few months to over five years.

The results of these tests are presented in this paper, along with a discussion of the methodology for testing and the significance of the tests.

Sample Preparation and Specimen Testing

Four sets of EPDM seam samples were fabricated in the laboratory in accordance with the manufacturers' instructions (Table 1). Four different EPDM manufacturers products were represented. Set 1C was fabricated using a 1.5 mm (60 mil) thick EPDM sheet and butyl based adhesive. Sets 2C and 3C were fabricated with 1.1 mm thick (45 mil) EPDM sheets and butyl based adhesives. Set 4C was a 1.1 mm thick (45 mil) EPDM sheet with a neoprene based adhesive.

To examine strength gain, the four sets were tested in T-peel at ages of 2 hours, 4 hours, 8 hours, 1 day, 3 days, and 7 days. Additional tests were conducted on sets 1C and 2C at 21 days. Preparation of the samples was performed under laboratory conditions of $23 \pm 2°$ C ($73 \pm 3°$ F) and 40 to 50 percent relative humidity. No specific attempts were made to control application rate of the adhesive in this study. Rollers were used to apply the adhesive and these were purchased from roofing suppliers and duplicated typical application practices in the area at the time of the study. Adhesive thickness was measured in each of the test samples by measuring the overall thickness of the lap and subtracting the measured thickness of the two overlapped sheets.

Since many precautions have been stated about the proper mixing of the adhesive prior to application [5], the adhesives were extensively stirred. Procedures for preparation of the splice area varied between manufacturers. Of the manufacturers' products tested, some required the use of a proprietary solvent for cleaning the splice area of the rubber, or use of a proprietary primer after solvent cleaning. Others only required cleaning the splice area with heptane, hexane, or unleaded gasoline. In these cases, unleaded gasoline was used to clean the splice area.

T-peel testing was conducted in general accordance with ASTM D 1876 [6]. The samples were prepared for T-peel testing by cutting 25 mm (1 in.) wide strips from the laboratory prepared seam in a direction perpendicular to the seam length. One end of the seam was peeled back about 25 mm (1 in.), and the opened portions of the seam were placed in the grips of a tensile testing machine. The load was then applied by pulling the lap in opposite directions, perpendicular to the adhesive line. The force required to separate the two layers of EPDM was measured and recorded continuously throughout the test. Since the force required to separate the sample varies slightly over the width of the seam, the force is averaged over the test length to obtain the average T-peel force per 25 mm (1 in.) of seam width.

Specific deviations from ASTM D 1876 were that the machine crosshead speed was 50 mm (2 in.) per minute, and the length of seam over which the T-peel strength was measured was about 50 to 76 mm (2 to 3 in.). This test speed has been widely used for the testing of EPDM seams. The length of the test area was less than specified in D 1876 because of limitations of the test apparatus.

Results and Discussion

<u>Laboratory Prepared Specimens - Cleaned</u>

Table 1 provides a listing of the EPDM sheet thickness, the adhesive type, the average adhesive thickness, and the T-peel test data for the laboratory prepared cleaned samples. Table 1 also lists the type of seam preparation normally prescribed by the manufacturer. The strength values shown in Table 1 represents an average of five peel tests conducted on strips cut from the laboratory test sample. The coefficient of variation for all of the laboratory prepared, cleaned samples was less than 20 percent. This data agreed with the precision stated in a previous study [4].

TABLE 1 -- Summary of Laboratory T-Peel Results - Cleaned

MEMBRANE DATA			
Set Number:	1C	2C	3C	4C
Sheet Thickness mm:	1.5	1.1	1.1	1.1
(mils):	(60)	(45)	(45)	(45)
Adhesive Type:	Butyl	Butyl	Butyl	Neoprene
Adhesive Thickness mm:	0.14	0.21	0.26	0.06
(mils):	(5.7)	(8.3)	(10.1)	(2.3)
Splice Preparation:	Splice Cleaner	Unleaded Gas + Primer	Unleaded Gas	Unleaded Gas + Primer

AGE	T-PEEL STRENGTH kN/m (lbf/in.)			
2 Hours	0.25 (1.40)	0.25 (1.40)	0.29 (1.66)	0.48 (2.72)
4 Hours	0.50 (2.86)	0.23 (1.34)	0.46 (2.60)	0.56 (3.18)
8 Hours	0.61 (3.48)	0.29 (1.64)	0.65 (3.74)	0.59 (3.38)
1 Day	0.78 (4.44)	0.50 (2.86)	0.89 (5.10)	0.52 (2.94)
3 Days	0.76 (4.34)	0.63 (3.58)	0.78 (4.44)	0.48 (2.76)
7 Days	0.94 (5.38)	0.70 (4.00)	1.16 (6.64)	0.64 (3.68)
21 Days	0.96 (5.50)	0.70 (4.02)	...	...

Best fit curves were plotted to describe the strength gain of sample sets 1C, 2C, and 3C (butyl based adhesives) and are shown in Figure 1. The strength development of sample set 4C (neoprene based adhesive) is represented in Figure 2. The curves were plotted using curve fitting computer software which described the best fit as a log function.

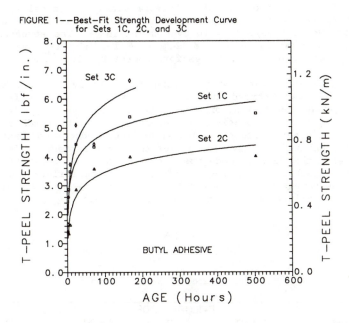

FIGURE 1—Best-Fit Strength Development Curve for Sets 1C, 2C, and 3C

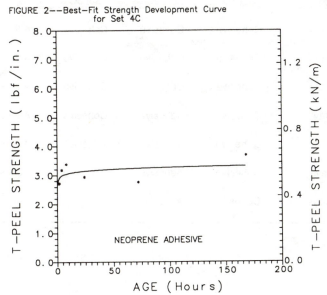

FIGURE 2—Best-Fit Strength Development Curve for Set 4C

For the butyl based adhesives, a comparison of the T-peel strengths at 3 days and 7 days to that at 21 days (sets 2C and 3C), indicates that about 75 percent of the strength gain occurred within 3 days, and over 85 percent of the strength gain occurred within 7 days. This represents peel strengths of about 0.6 to 0.9 kN/m (3.3 to 5.5 lbf/in.) in 3 days, and about 0.7 to 1.1 kN/m (3.8 to 6.4 lbf/in.) in 7 days. The ultimate strength (considered to be about 21 days for the purposes of this study), was on the order of 0.8 to over 1.0 kN/m (4.4 to over 6 lbf/in.). The strength of the butyl based adhesive at 0.8 kN/m (4.4 lbf/in.) is lower than that reported in previous studies for laboratory cleaned samples.

The mode of failure, whether cohesive, adhesive, or a mixture of the two, was observed during peel testing. Cohesive failure occurred when separation of the EPDM strips occurred in the adhesive itself, and adhesive was observed to be still adhered to both of the sheet surfaces. Adhesive failure occurred when the adhesive separated cleanly from one of the sheet surfaces. In some sample sets, the failure mode was observed to be a combination of cohesive and adhesive failure. Table 2 lists the failure mode for each of the sample sets at each test age. When the mode was mixed, the primary failure type is indicated in the table.

TABLE 2 -- Primary Failure Mode for Laboratory Samples - Cleaned

FAILURE MODE			
Age	Set 1C	Set 2C	Set 3C	Set 4C
2 Hours	Cohesive	Cohesive	Cohesive	Adhesive
4 Hours	Cohesive	Cohesive	Cohesive	Adhesive
8 Hours	Cohesive	Cohesive	Cohesive	Adhesive
1 Day	Cohesive	Adhesive	Cohesive	Adhesive
3 Days	Cohesive	Adhesive	Adhesive	Adhesive
7 Days	Cohesive	Adhesive	Adhesive	Adhesive
21 Days	Cohesive	Adhesive	...	...

The primary mode of failure for the butyl based adhesives (Sets 1C, 2C, and 3C) was cohesive, at least up to about one to three days. At that point, Sets 2C and 3C failed adhesively. As the T-peel strength reached about the 0.5 to 0.8 kN/m (2.9 to 4.6 lbf/in.) level, the failure mode changed from cohesive to adhesive. This was unexpected. A well bonded seam would be expected to fail cohesively at later ages, such as in Set 1C. In spite of cleaning the rubber sheet, the failure occurred due to an interfacial effect. This was also observed in a study by Rossiter et al [7]. In this study, even cleaned specimens failed adhesively at later ages (7 to 14 days) at strengths of about 0.7 to 1.0 kN/m (4 to 6 lbf/in.). The combination of lower than expected strength (Set 2C), and adhesive failure implies that weakest part of the seam was at the interface of the rubber and the adhesive.

With regard to the neoprene adhesive, its initial strength is greater than that of the butyl based adhesives (Table 1). However, the point at which the neoprene based adhesive reached 85 percent of its 7-day strength was about 4 to 8 hours. The ultimate strength of the neoprene based adhesive (considered to be at about 7 days for the purposes of this study), was about 0.6 kN/m (3.3 lbf/in.). Based on interpolation of the best-fit curve, the neoprene based adhesive peel strength was only about 50 to 75 percent of that achieved using a butyl based adhesive. The neoprene adhesive had much higher initial strength than the butyl, however, little increase in strength was observed after about 8 hours. For the neoprene based adhesive, the failure mode was adhesive in all cases.

Laboratory Prepared Lap Seams - Uncleaned

The materials used for preparing the uncleaned specimens were from the same manufacturers and using the same sheet thicknesses as in sets 1C and 2C. In lieu of the recommended cleaning process, the splice area of the rubber sheet was simply wiped with a clean dry cloth to remove any gross particulate contamination. Following this minimal preparation, the adhesive was applied and the seams constructed in accordance with the manufacturers' recommendations.

The uncleaned lap seams were tested at the ages of 4 hours, 8 hours, one day, 3 days, and 7 days. Table 3 provides a listing of the sheet thickness, adhesive type, and a summary of the T-peel test data. The strength values shown in Table 3 represents an average of five peel tests conducted on strips cut from the laboratory test sample. The coefficient of variation for all of the laboratory prepared, uncleaned samples was less than 20 percent. Table 4 shows the primary failure mode for each set of test strips at each test age.

TABLE 3 -- Summary of Laboratory Test Results - Uncleaned

MEMBRANE DATA		
Set Number:	1 UC	2 UC
Sheet Thickness mm:	1.5	1.1
(mils):	(60)	(45)
Adhesive Type:	Butyl	Butyl
Adhesive Thickness mm:	0.15	0.26
(mils):	(5.8)	(10.2)
Splice Preparation:	None	None
AGE	**T–PEEL STRENGTH**	**kN/m (lbf/in.)**
4 Hours	0.70 (3.98)	0.36 (2.06)
8 Hours	0.71 (4.08)	0.43 (2.44)
1 Day	0.87 (4.96)	0.75 (4.26)
3 Days	0.78 (4.48)	0.65 (3.70)
7 Days	0.82 (4.68)	0.54 (3.08)

TABLE 4 -- Primary Failure Mode for Laboratory
Samples - Uncleaned

FAILURE MODE		
AGE	Set 1 UC	Set 2 UC
4 Hours	Cohesive	Cohesive
8 Hours	Cohesive	Adhesive
1 Day	Adhesive	Adhesive
3 Days	Adhesive	Adhesive
7 Days	Adhesive	Adhesive

The strength development of sample sets 1UC and 2UC is shown in Figure 3, and is represented by a series of best-fit curves. Figure 3 also shows the strength data in relation to that of the cleaned samples (sets 1C and 2C).

FIGURE 3—Best-Fit Curve for Sets 1UC and 2UC
and the Corresponding Cleaned Samples

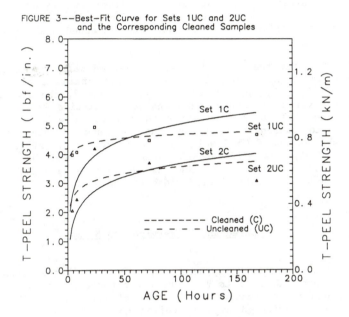

The ultimate strength of the uncleaned samples was about 0.8 kN/m (4.8 lbf/in.) in the case of set 1UC, and about 0.6 kN/m (3.7 lbf/in.) in the case of set 2UC. The early age specimens (4 to 8 hours) failed in the cohesive mode. As the adhesive gained strength, the surface effects of the uncleaned rubber sheets controlled the failure mode and the specimens failed adhesively, as expected. The strength of set 1UC appeared to be somewhat higher than observed in previous studies for an as-received uncleaned sample. Martin et al [8] showed that in general, as the level of contamination decreased, the peel strength increased. Wiping gross particulate contamination from the rubber surface may have been effective in removing some of the release agent, and achieving artificially high uncleaned peel strength.

In comparing the cleaned versus uncleaned data for set 1, the observed strength decrease due to lack of cleaning was about 15 percent. This difference was much less than expected based on data from Rossiter et al [7], and Martin et al [8]. Since these studies have shown butyl based adhesives to achieve strengths of 1.2 to 1.6 kN/m (7 to 9 lbf/in.), the small difference in cleaned versus uncleaned strength may be due to the relatively low strength of the cleaned specimens. Since the cleaned set (1C) consistently failed cohesively, and the uncleaned set (1UC) failed adhesively after 8 hours, the strength reduction appears to be an effect of cleaning.

In the case of set 2, the reduction in strength due to lack of cleaning was less than 10 percent, which is considerably less than expected based on previous studies [3,7,8]. However in this case, the difference may not have been expected to have been great, based on the observation that both sets failed adhesively, and the cleaned laboratory strength of set 2C was lower than expected. The adhesive failure mode indicates that an interfacial effect controlled the failure mode in both the cleaned and uncleaned samples.

The data for the cleaned versus uncleaned sets were compared at the 0.05 significance level using the statistical t-test technique [10]. In the case of both sets 1 and 2, the average of the cleaned versus uncleaned specimens was statistically different, showing that although the strength levels were closer than expected, the strength reduction due to lack of cleaning was significant at the 0.05 level. As we increase the significance level to 0.01, the difference between sets 2C and 2UC is not statistically different. In the case of set 1C and 1UC, there still remains a statistical difference at the 0.01 level.

Analysis of Field Prepared and In-service Lap Seams

Field prepared EPDM lap seam samples were collected from five different roofs (Sets F1 to F5). The first three roofs (Sets F1 to F3) were under construction at the time of the study. These lap samples were fabricated by the roofing contractor during routine construction. The samples are intended to represent typical field seaming practice. The samples were fabricated by the contractor from pieces of membrane measuring about 0.3 x 1 m (1 x 3 ft.). In all cases, the seaming practices of the contractor appeared to be in compliance with the manufacturers' instructions.

Rate of application of the adhesive was not measured for sets F1, F2, and F3. Sets F4 and F5 were taken from roofs in-service and installation was not observed. The thickness of the adhesive for each of the test strips was measured in the lab (Tables 5 and 6). For sets F1 through F3, the samples were returned from the field the same day they were formed. The age of the seams when tested ranged from about one month to several months. Based on the strength gain data presented in this study (Figures 1 and 2), the strength of the field specimens was assumed to have stabilized at its maximum level by the time the seams were tested.

The field lap specimens for sets F1 and F2 were fabricated using materials from the same manufacturer as sets 1C and 1UC. The field lap for set F3 was fabricated using materials from the same manufacturer as sets 2C and 2UC. A description of the materials, the temperature and weather conditions under which they were constructed, and the T-peel test data are shown in Table 5. The average T-peel strength shown in Table 5 represents an average of ten strength tests conducted on strips cut from the field samples.

TABLE 5 -- Summary of Field Prepared Seams - Roofs Under Construction

MEMBRANE DATA			
Set Number:	F1	F2	F3
Sheet Thickness mm:	1.5	1.5	1.1
(mils):	(60)	(60)	(45)
Similar Set:	1C	1C	2C
Adhesive Type:	Butyl	Butyl	Butyl
Adhesive Thickness mm:	0.13	0.22	0.08
(mils):	(5.1)	(8.7)	(3.2)
Age When Tested:	1 month	1 month	8 months
APPLICATION CONDITIONS			
Season:	Summer	Summer	Winter
Temperature (°C):	30	30	0
Weather:	Sunny	Sunny	Cloudy
T-PEEL TEST DATA			
Avg. T-Peel Strength kN/m:	1.08	0.97	0.37
(lbf/in.):	(6.18)	(5.57)	(2.14)
Range kN/m:	1.02 – 1.19	0.87 – 1.07	0.32 – 0.44
(lbf/in.):	(5.8 – 6.8)	(5.0 – 6.1)	(1.8 – 2.5)
Std. Deviation kN/m:	0.07	0.05	0.05
(lbf/in.):	(0.40)	(0.31)	(0.26)
C.O.V. (%):	6.5	5.5	12.1
Primary Failure Mode:	Cohesive	Cohesive	Adhesive

In the case of sets F4 and F5, the seams were performing satisfactorily with no reported leaks in both cases. The manufacturer of set F4 was the same as that of set 3C of the cleaned laboratory samples, and the roof was constructed in a loose laid, ballasted configuration. The manufacturer of set F5 was not determined, and the construction history of this roof was not available at the time of the study. However the adhesive was identified as neoprene.

A summary of the in-service lap seams and the T-peel test data is shown in Table 6. The average T-peel strength shown in Table 6 represents tests conducted on strips cut from each of two or three different areas of the roof (two areas for set F4 and three areas for set F5). Five T-peel tests were conducted for each test area, and the average shown in Table 6 is a combined overall average.

TABLE 6 -- Summary of In-Service Lap Seams - Existing Roofs

MEMBRANE DATA		
Set Number:	F4	F5
Sheet Thickness (mm):	1.1	1.5
(mils):	(45)	(60)
Adhesive Type:	Butyl	Neoprene
Adhesive Thickness mm:	0.17	0.13
(mils):	(6.8)	(5.3)
Age When Tested:	6 months	5 years
T–PEEL TEST DATA		
Average T–Peel Strength kN/m:	0.58	0.16
(lbf/in.):	(3.31)	(0.93)
Range kN/m:	0.39 – 0.77	0.07 – 0.35
(lbf/in.):	(2.2 – 4.4)	(0.4 – 2.0)
Std. Deviation kN/m:	0.16	0.08
(lbf/in.):	(0.89)	(0.43)
C.O.V. (%):	26.9	46.6
Primary Failure Mode:	Adhesive	Adhesive

Figure 4 is a graphic representation of the average T-peel strength of all of the field samples. The figure also includes the comparable (cleaned) laboratory strengths at 7 days for the same manufacturer (except for set F5).

FIGURE 4 -- In-Service and Field Seams
(Avg. Field vs Avg. 7-Day Lab Strength)

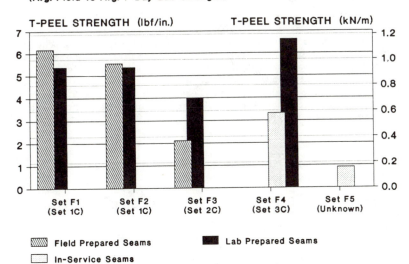

In the first two cases (sets F1 and F2 versus 1C), the contractor prepared samples had peel strengths comparable to that achieved in the laboratory. The peel strength for set F3 was about half of the peel strength achieved in the laboratory sample (2C). The failure mode in peel for sets F3 and 2C were both adhesive, indicating that failure was due to surface effects. Reasons for the apparent low field strength of set F3 and the failure mode were not investigated in this study. However, it is interesting to note the differences in application conditions for sets F1 and F2 versus those for set F3 (Table 5). Other factors, such as temperature and humidity conditions may have been responsible for differences in peel strength.

In the case of sample set F4 versus set 3C, the average peel strength of the specimens cut from the roof was only about half of the strength achieved in the laboratory samples prepared with clean rubber. Since this roof had been in-service approximately 6 months and seam construction was not observed, reasons for the relative low strength of the field sample were not known. The failure mode for set F4 again indicates a surface effect, but the factors leading to the adhesive failure mode were not determined.

In the case of set F5, the average T-peel strength was less than 0.2 kN/m (1.0 lbf/in.). As indicated, this adhesive was identified as neoprene based, and the seams had been performing satisfactorily for over five years. This observation suggests that lap seams are capable of performing satisfactorily at relatively low peel strengths.

Effect of Adhesive Thickness

An important factor in making field splices with EPDM sheets is the coverage rate of the adhesive. Although this factor was not controlled in this study, adhesive thickness of all of the test specimens was measured and recorded in the tables. Another study by Watanabe [11] has shown that adhesive thickness effects T-peel strength. The study shows that the greater the thickness (up to a maximum level), the higher the peel strength. Figure 5 presents a plot of the average adhesive thickness versus T-peel strength for all of the samples in the study.

FIGURE 5 -- T-Peel Strength Versus
Adhesive Thickness for All Specimens

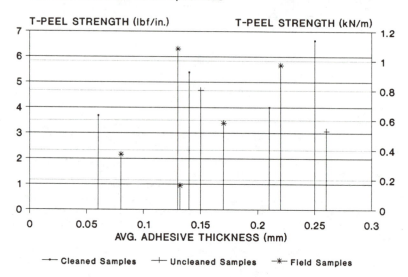

Not unexpectedly, no thickness effect was found. The lack of correlation between strength and thickness can be attributed to the numerous other parameters that have an effect on peel strength but were not controlled in this study. Further testing to isolate the adhesive thickness as a test parameter would be required to show statistically significant correlations between thickness and T-peel strength.

Summary

As previous studies and field experience have shown, there are a number of factors which influence the integrity of a field fabricated EPDM lap seam. In performing a laboratory and field study, it would be an enormous task to separate and evaluate each of these variables. The purpose of this study was to look at some membranes in the laboratory, under construction, and in-service, in order to characterize the peel strength achieved by typical seaming practice. The purpose of the laboratory testing in this case was not to isolate a number of test variables and study their effect, but to start establishing values of seam strength that could be expected under reasonably careful preparation, and compare their peel strengths to those from a limited number of field samples.

In summary, a number of observations can be made from this study.

o The strength gain characteristics of a butyl versus a neoprene based adhesive are significantly different in that the ultimate T-peel strength of the neoprene is only about 1/2 to 3/4 that of the butyl. However, the neoprene adhesive gains strength much more rapidly at early ages.

o The ultimate strengths of the butyl adhesives tested in this study were somewhat lower than expected and as a result, the differences between the cleaned and uncleaned laboratory samples were smaller than previously reported. Surface effects played an important role in controlling the failure mode in the laboratory cleaned and uncleaned samples. Whereas we would expect to see cleaned samples fail in the cohesive mode in all cases, this study showed some cleaned samples failing in the adhesive mode.

o In the case of the field samples tested in this study, three observations can be made.

1. In two cases (F1 and F2), the strengths achieved by the contractor in the field were equivalent to those achieved in the laboratory.

2. In two other cases (F3 and F4), the strengths achieved in the field were only about 1/2 of the laboratory strengths from the comparable samples. The specific reasons for the relatively low field strength requires further study.

3. Although the T-peel strength is relatively low (sets F4 and F5), the seams in these cases were performing well.

o The effect of coverage rate and adhesive thickness may be an important factor in determining seam strength, however, further study is required.

REFERENCES

[1] Cullen, William C., "Project Pinpoint analysis: trends and problems in low-slope roofing 1983-1988", NRCA

[2] Rossiter, Walter, J., Jr., "The Effect of Application Parameters on Adhesive-Bonded Seams in Single-Ply Membranes," Proceedings, Second International Symposium on Roofing Technology, NRCA, pp. 383-390, September 1985

[3] Rossiter, Walter J. Jr., "Further Investigation of the Effect of Application Parameters on Adhesive-Bonded Seams of Single-Ply Roof Membrane," Materials and Structures, Volume 21, pp. 243-249, 1987

[4] Westley, S.A., "Bonding Ethylene Propylene Diene Monomer Roofing Membranes: The Theory and Practice of Adhering Vulcanized Ethylene Propylene Diene to Itself," Single-Ply Roofing Technology, ASTM Special Technical Publication 790, pp. 90-108, ASTM, 1981

[5] Russo, Michael, "Single-Ply Seaming: The State of the Art," RSI, August 1988

[6] ASTM D 1876, "Standard Test Method for Peel Resistance of Adhesives (T-Peel Test), " Annual Book of ASTM Standards, Volume 15.06, ASTM, 1983

[7] Rossiter, Walter, J. Jr., Seiler, James F. Jr., Stutzman, Paul E., "Field Testing of Adhesive-Bonded Seams of Rubber Roofing Membranes," Proceedings of the 9th Conference on Roofing Technology, NIST/NRCA, pp. 78-87, May 1989

[8] Martin, Johnathan W., Embree, Edward, Rossiter, Walter J. Jr., "Effect of Contamination Level on Strength of Butyl-Adhered EPDM Joints in EPDM Single-ply Roofing Membranes," Proceedings of the 9th Conference on Roofing Technology, NIST/NRCA, pp. 64-72, May 1989

[9] Chmiel, Chester T.,"History and Development of EPDM Splice Adhesives," RSI, November 1987

[10] Bhattacharyya, Gouri K., Johnson, Richard A., "Statistical Concepts and Methods," John Wiley & Sons, Inc., 1977

[11] Watanabe, H., Rossiter, Walter J. Jr., "Effects of Adhesive Thickness and Open Time on the Peel Strengths of Adhesive Bonded Seams of EPDM Rubber Roof Membranes," Roofing Research and Standards Development: 2nd Volume, ASTM STP 1088, American Society for Testing and Materials, 1990

Hiroshi Watanabe and Walter J. Rossiter, Jr.[1]

EFFECTS OF ADHESIVE THICKNESS, OPEN TIME, AND SURFACE CLEANNESS
ON THE PEEL STRENGTH OF ADHESIVE-BONDED SEAMS
OF EPDM RUBBER ROOFING MEMBRANE

REFERENCE: Watanabe, H. and Rossiter, W. J., Jr., "Effects
of Adhesive Thickness, Open Time, and Surface Cleanness on
the Peel Strength of Adhesive-Bonded Seams of EPDM Rubber
Roofing Membrane, "Roofing Research and Standards
Development: 2nd Volume, ASTM STP 1088, Thomas J. Wallace
and Walter J. Rossiter, Jr., Eds., American Society for
Testing and Materials, Philadelphia, 1990.

ABSTRACT: A laboratory study was conducted to examine the
effects of adhesive thickness, open time, and surface
cleanness on T-peel strength of EPDM (ethylene propylene
diene terpolymer) rubber seam specimens. Seam specimens
bonded with butyl-based contact adhesive were tested after a
2-week cure time. The peel strength generally showed a
positive dependency on the adhesive thickness except that it
tended towards a plateau value for thick adhesive layers.
The leveling of peel strength at large thickness might result
from the presence of small voids in the adhesive layer. The
peel strength was not dependent on the open times used in
this study. Increased levels of surface contamination
lowered the peel strength and changed the failure mode from
cohesive to adhesive.

KEY WORDS: adhesive thickness, contamination, EPDM rubber,
open time, peel test, roofing, seams, single-ply membrane

Single-ply membranes have come into wide use as the waterproofing
component of low-sloped roofing systems in the United States and other
countries such as Japan. EPDM (ethylene propylene diene terpolymer)
rubber currently holds a major portion of the single-ply market. A
significant factor in EPDM's popularity is that the performance of
these systems have been generally satisfactory in the field. However,
as is the case of roofing systems in general, performance has not been

---

[1]Mr. Watanabe is a Guest Researcher (from Takenaka Corp., Tokyo,
Japan), Dr. Rossiter is a Research Chemist. Both are at the Building
Materials Division, Center for Building Technology, National Institute
of Standards and Technology, Gaithersburg, MD 20899.

without concern. A critical factor of any roofing system is the
ability of the seams to remain watertight over the service life of the
roof. The weather resistance of EPDM membranes has been satisfactory
[1,2], a fact attributed to the general chemical inertness and non-
polar nature of the rubber. This same chemical inertness makes
adhesive-bonding of the sheets difficult. A further complication is
that the seams are fabricated in the field using manual techniques
under a variety of weather conditions.

Methods are needed to provide guidelines for the formation of seams
to assure their quality, not only for vulcanized rubber materials but
also for other single-ply products. ASTM (American Society for Testing
and Materials) Subcommittee D08.18 has established a task group
addressing seam performance. Data on the characterization of newly-
formed and aged systems are needed to support the work of the task
group.

Several studies [3-10] have been conducted on EPDM systems to
provide baseline data on the factors affecting seam performance. These
studies have addressed both short-term bond strength tests and long-
term creep rupture tests. Factors addressed in these studies have
included temperature, stress level, rate of loading, voids in the
adhesive layer, pressure during application, and contamination of the
rubber surface. One important finding was that lap-shear tests were
not sensitive to many factors which would be expected to affect bond
strength adversely [5]. Conversely, T-peel tests were sensitive to
such factors, particularly surface contamination [8]. Consequently, a
recommendation [11] was made that T-peel testing should provide the
basis of a field method for the quality assurance of newly-formed
seams.

The present study aims at providing additional insight into the
factors affecting the results of T-peel tests, which is necessary to
develop a quality assurance methodology. A major factor addressed is
the effect of the thickness of the adhesive layer on the peel strength
of the joint specimens. This factor has not been addressed in previous
studies of EPDM seams. From a theoretical viewpoint, peel strength
should increase with an increase in adhesive thickness [12,13].
Typical adhesive thickness of seams sampled in the field ranged from
approximately 0.1 to 0.4 mm (0.004 to 0.016 in.)[2].

A second factor investigated is adhesive open time, which is
defined as the time between applying the adhesive to the sheets and
joining them to form the bond. Open time allows the solvents present
in the contact adhesive to evaporate before bonding. However, it may
be postulated that excessively long open times are detrimental to
satisfactory bonding due to loss of tack. In practice, open times
generally range from 10 to 30 min.

The final factor included in the study is the surface cleanness of
the rubber sheets used to form the seam specimens. The surfaces were
contaminated with talc-like particles used as a release agent in the
production of EPDM rubber roofing materials. Previous laboratory

[2]NIST unpublished data.

studies [8-10] have shown that surface contaminants may significantly reduce peel strength. In addition, some seam samples removed from roofs have displayed comparatively low values of peel strength and also platelet particles, indicative of release agent, on the surface of the rubber [10]. Surface cleanness was included in the present study because it was of interest to investigate the combined effects of this factor with adhesive thickness and open time.

EXPERIMENTAL

Materials

The specimens were fabricated using a commercially available EPDM roofing membrane and a butyl-based contact adhesive supplied by the manufacturer of the membrane. The membrane material had a nominal thickness of 1.5 mm (0.060 in.), was non-reinforced, and had a talc-like release agent on its surfaces. The elastic moduli of the membrane were 1.17 MPa (170 lbf/in.^{2}) at 50% elongation and 1.21 MPa (176 lbf/in.^{2}) at 100% elongation. The density of the EPDM rubber was 1.16 g/cm^3 and that of the adhesive was 0.86 g/cm^3 (when taken from a well-stirred container). The adhesive was kept in small, closed containers until use and was thoroughly stirred before application.

Application Factors

The three factors included in the study as experimental variables are given in Table 1. Adhesive thickness had four nominal levels representing the measured thickness after a 2-week cure. Open time ranged from 5 min to 15 h. Surface cleanness had three nominal levels, quantified according to the grayscale measurement technique presented by Martin et al. [9]. The nominal grayscale levels were 30, 90, and 160, designated as clean, lightly contaminated, and heavily contaminated, respectively.

TABLE 1 -- Factors included in the study

Factor	Level
Nominal adhesive thickness[a]	0.10, 0.17, 0.36, 0.54 mm (0.004, 0.007, 0.014, 0.021 in.)
Open time	Range: 5 min to 15 h
Surface cleanness	Three nominal grayscale levels[b]: 30 = clean 90 = lightly contaminated 160 = heavily contaminated

[a] Based on the measured thickness of the specimens after a 2-week cure.
[b] According to the method of Martin et al. [9].

Specimen Preparation

The preparation and curing of the seam specimens were carried out during the summer at ambient laboratory conditions, approximately 23 °C (73 °F) and between 60 and 68% RH. The EPDM membrane sheet was cut into strips, having dimensions of 25 by 125 mm (1 by 5 in.) with a 75-mm-long (3 in.) section at the strip end used for adhesive bonding. The strips were initially cleaned by scrubbing both sides with soap and water, which was followed by rinsing with water. The strips were dried at ambient laboratory conditions and were further scrubbed with a clean cloth saturated with heptane. For each set of experimental conditions, three replicates were prepared.

After the strips were cleaned, they were matched into pairs (side A and side B) for fabricating the T-peel specimen. As for the clean specimens (grayscale 30), neither side A nor side B was contaminated. As for the contaminated specimens, side A strips were kept clean, while side B strips were contaminated to their specified levels. The contamination procedure has been described by Martin et al. [9]. It consists of passing the strips under a uniform spray of talc-like particles suspended in heptane. The amount of particles deposited (i.e., degree of contamination) depends on the number of passes under the atomizer.

The adhesive was applied to the strips using a knife-coat[3] method [14] to provide uniformity of the adhesive layer at the desired nominal thickness. Four rubber strips were secured on a vacuum table side by side between two flat metal plates (about 200 by 25 mm or 8 by 1 in.). The adhesive was then poured on one end of the strips and a draw-down blade was pulled along the length of the strips. Four draw-down blades were selected to provide wet film thicknesses of about 0.3, 0.6, 1.1, and 1.9 mm (0.01, 0.02, 0.04, 0.07 in.). These wet film thicknesses yielded the nominal thickness levels in Table 1. After a predetermined open time, the adhesive-covered surfaces of the pairs of strips were joined. Within 1 min after joining, the seam specimens were pressed at 0.69 MPa (100 lbf/in.^{2}) for 5 to 10 s with a laboratory press. The seam specimens were then cured for 2 weeks before peel testing.

Measurement of adhesive thickness: The thickness of the adhesive was determined by subtracting the thicknesses of the two strips from the total thickness of the completed seam specimen. Measurements were generally made after the specimens had cured for 2 weeks, although a series of measurements was made at cure times ranging from 1 min to 2 weeks. The adhesive thickness of each strip or completed membrane specimen was taken as the average of measurements at three locations along the center axis of the strip: about 25 mm (1 in.) from each end and at the midpoint within the bond area. The result obtained using this procedure was that, for an individual specimen, the maximum range of the three thickness calculations was less than 10% of the average. For the four sets of specimens having the assigned thickness levels

[3]This method uses an adjustable knife blade (draw-down blade), bar, or rod to control dispersion of the adhesive on the strip [14]. The adhesive thickness is controlled by the distance between the blade edge and the substrate surface.

given in Table 1, the average calculated thickness were 0.099, 0.168, 0.356, and 0.538 mm (0.004, 0.007, 0.014, and 0.021 in.), and the standard deviations were 0.010, 0.028, 0.020, and 0.054 mm (0.0004, 0.0012, 0.0008, and 0.002 in.), respectively.

Measurement of surface cleanness: The technique described by Martin et al. [9] was used to quantify the cleanness of the membrane strips or the extent of contamination resulting from deposition of the talc-like particles. In brief, the technique uses computer-image processing to measure the reflectance of tungsten light from the surfaces of the membrane strips. Light reflectance is directly proportional to the level of contamination. Reflectance is quantified according to a grayscale output from an image processor. The grayscale value is zero for an absolute black surface and 256 for an absolute white surface.

The reflectance of each strip was measured at 20 locations within the bond area of a T-peel specimen. The average and standard deviation values were computed for each grayscale level and are given in Table 2. Because the standard deviations were small (coefficient of variation less than 9%), the variations in surface cleanness were considered to be negligible. Thus, the peel data discussed in the report are analyzed as a function of nominal grayscale level and not of measured values.

TABLE 2 -- Measured grayscale values of the membrane strips used to prepare the seam specimens

Nominal Grayscale Level	Measured Grayscale Value			
	Side A		Side B	
	Average	S.D.[a]	Average	S.D.[a]
30	31.6	1.4	31.5	1.4
90	30.0	2.7	93.2	2.6
160	30.4	2.7	160.2	2.4

[a] Standard deviation.

T-Peel Test

The test procedure followed ASTM Test Method for Peel Resistance of Adhesives (T-Peel Test) (D1876), except that the bond length was 75 mm (3 in.) and the load was applied at a rate of 50 mm/min. (2 in./min). The tests were performed at ambient laboratory temperature (approximately 23 °C or 73 °F).

The locus of failure was determined by visual inspection. After the test, the mass of adhesive remaining on each strip was determined by subtracting the initial mass. The mass percentage of adhesive remaining on the side B rubber was calculated to characterize the failure mode.

RESULTS AND DISCUSSION

Effect of Open Time and Cure Time on Adhesive Thickness

From previous tests[4], it was known that adhesive-bonded seam specimens continued to lose mass for at least a week after preparation. For the present study, it was required to have seam specimens that had attained relatively constant thickness. A preliminary experiment was conducted to determine the change in thickness as a function of cure time, open time, and thickness of the adhesive layer.

Typical results are presented in Fig. 1, whereby two sets of curves are given: one for clean specimens having a nominal adhesive thickness of 0.54 mm (0.021 in.), and the other with that of 0.10 mm (0.004 in.). These two values bracketed the nominal thicknesses used in the study. For both sets of curves and for a given open time, the adhesive thickness was found to decrease with cure time. In general, the greatest decrease occurred during the first day of cure. The thickness

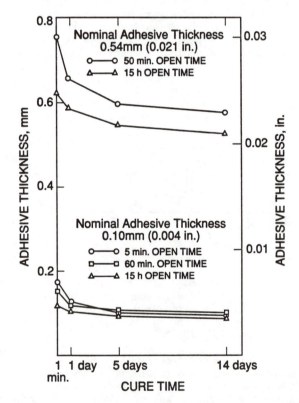

FIG. 1 -- Adhesive thickness as related to open time for various cure times of seam specimens.

[4]NIST unpublished data.

of the specimens was still decreasing between 5 and 14 days of cure. However, the changes were considered to be small. Consequently, it was decided to use a 2-week cure time in further experimentation.

A final point for discussion is the relation between open times used in the preliminary experiment and those employed in practice. In the preliminary experiment, the open time varied from a few minutes to several hours (Fig. 1). In practice, the open time is a point at which the roofing mechanic judges that sufficient solvent has evaporated from the freshly applied adhesive to allow seam formation. The judgment is generally based on two hand actions (finger criteria) performed by the mechanic. First, the surface of the adhesive is lightly touched with a finger to determine whether the adhesive appears to be dry[5]. Second, the finger is pressed into the bulk of the adhesive to check whether solvent is present and whether the adhesive bulk resists flow when pushed with the finger. If the mechanic finds that the adhesive surface is dry to the touch and the bulk resists flow, then the open time is considered sufficient for bonding.

In the preliminary experiment, judgments of dryness and bulk flow resistance were made using the finger criteria on freshly applied adhesives. The thicknesses of the fresh adhesives varied and represented the four nominal adhesive levels of the cured specimens (Table 1). Because the loss of solvent from the applied adhesive varies with the initial thickness, it was expected that the estimated times to achieve loss of dryness to the touch and resistance to bulk flow would vary with thickness. As given in Table 3, this was found to be the case: greater adhesive thickness corresponded to longer time to attain dryness to the touch and resistance to bulk flow. It is cautioned that the values given in Table 3 represent subjective judgment under the given environmental conditions of the laboratory.

TABLE 3 -- Open times estimated for the four nominal adhesive thicknesses as determined according to common roofing practice[a,b]

Nominal Adhesive Thickness		Dry-to-touch[c]	Bulk Flow Resistance[d]
mm	(in.)	min	min
0.10	(0.004)	10	30
0.17	(0.007)	20	40
0.36	(0.014)	30	50
0.54	(0.021)	50	75

[a] See text for description.
[b] The values given in the Table are subjective. Another person under different environmental conditions might report other values.
[c] The time at which the adhesive surface appears to be dry when touched with a finger.
[d] The time at which the bulk of the adhesive resists flow when pushed with a finger.

[5]In contact adhesive use, this is often referred to as "dry to the touch" [15].

Effect of Open Time on Peel Strength

Peel strength was determined for seam specimens of various open times for each of the four nominal adhesive thicknesses used in the study (Fig. 2). All specimens were prepared from cleaned rubber sheets, and tested after a 2-week cure at room temperature. The results indicated that the open time did not appear to have a major effect on the strength. In contrast, strength did increase with an increase in nominal adhesive thickness.

Analysis of variance was performed to examine whether, for a given thickness level, the peel strength differed with a change in open time. The results of this analysis were: (1) for the nominal thickness levels of 0.10 and 0.36 mm (0.004 and 0.014 in.), no significant differences were found, and (2) for the nominal thickness levels of 0.17 and 0.54 mm (0.007 and 0.021 in.), significant differences between strength for various open times were present. However, in the latter case, the scatter in the peel strength values for the various open times was random, and no trends in the data were found. Consequently, no practical significance was attributed to these analysis of variance results. It was not considered probable that the strength values as a function of open time would be constant only at the two nominal thickness levels. It was believed that the major variations in strength for specific nominal thickness levels were caused by uncontrolled and undefined differences in specimen preparation. For instance, the scatter in the data for the thickest adhesive may be influenced by voids in the adhesive layer formed by solvent entrapment (see discussion below).

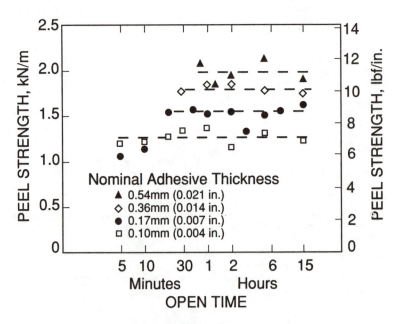

FIG.2 -- Peel strength versus open time at four levels of nominal adhesive thickness. (The open time axis is logarithmic.)

Effect of Adhesive Thickness and Surface Cleanness

   Peel strength: Peel strength was determined for specimens having
varying adhesive thickness (corresponding approximately to the nominal
levels in Table 1) and three levels of surface cleanness. In addition,
two open times which varied with nominal adhesive thickness were used:
dry-to-touch open time and a bulk flow resistance open time as
discussed above and given in Table 3. The results of the tests are
presented in Fig. 3. Effects were found due to adhesive thickness and
surface cleanness, whereas open time had no effect. In the latter
case, the finding was not unexpected, based on the previous results
given in Fig. 2. Statistical analysis (t-test) of the open time data
in Fig. 3 indicated that, for all pairs of open time data at the given
thickness and grayscale levels, no differences in average strength were
found at the 0.05 significance level.

   As is evident in Fig. 3, for a given adhesive thickness, the peel
strength of the specimens decreased with increasing contamination.
Similar findings having been reported by Martin et al. [9], and by
Rossiter and Seiler [10]. In the present study, for a given level of
surface contamination, the peel strength generally increased with an
increase in adhesive thickness, with the clean specimens apparently
reaching a plateau value at a thickness of about 0.4 mm (0.016 in.).

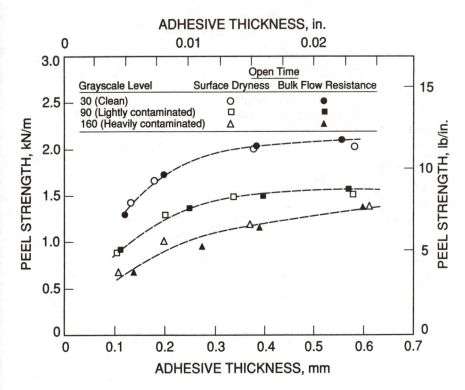

FIG. 3 -- Peel strength versus adhesive thickness at three
grayscale levels of specimen surface.

In all cases, the maximum strength was found for specimens that had
the thicker adhesive layers and cleaned rubber surfaces. Conversely,
the minimum strength values were observed for the thinnest and most
contaminated specimens. Specimens which were clean and had thin
adhesive layers displayed comparable peel strengths to those that were
contaminated and had thick layers of adhesive. This latter finding
implies that increasing the amount of adhesive in the specimens can, to
a degree, overcome the adverse effect of surface contamination.
Therefore, if the T-peel test is used to determine a minimum acceptable
bond strength for quality control purposes, the thickness of the
adhesive layer in the specimen should also be reported.

Failure mode during peel: Fig. 4 gives the mass percentage of
adhesive which remained on the side B rubber strip after the peel test.
These data may be interpreted on the following assumptions, which were
previously discussed by Martin et al. [9]:
1. If the failure mode was cohesive, the adhesive would be expected
   to be evenly distributed on the two strips.
2. If the failure mode was adhesive, the adhesive might be expected
   to remain on the uncontaminated (side A) strip. In the present
   study, the locus of adhesive failure was expected to be between
   the adhesive and the surface of the side B strip where the talc-
   like contaminant was applied.

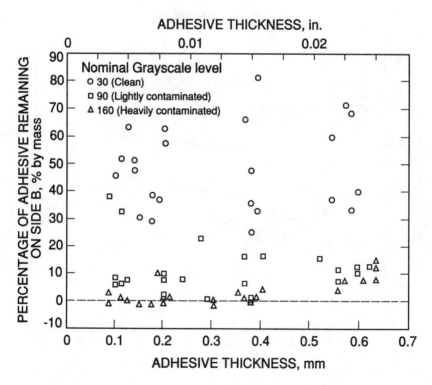

FIG. 4 -- Dependance of failure mode upon surface cleanness
and adhesive thickness.

The data in Fig. 4 are consistent with these expectations, and supported by observations noted during peel testing. The failure modes for the specimens are summarized as follows:

1. The clean specimens failed cohesively, although the distribution of the adhesive on the side B ranged from about 25 to 80%. It was observed that, for some specimens, the locus of failure occurred closer to one strip than the other, and not along the mid-section of the adhesive layer.

2. The lightly contaminated specimens underwent a mixture of adhesive and cohesive failure. The percent adhesive on the side B strip varied between about 0 and 30%.

3. The heavily contaminated specimens essentially displayed adhesive failure if the adhesive thickness was about 0.4 mm (0.016 in.) or less. In these cases, the mass percent adhesive on the side B strip was 4% or less. The thickest specimens (nominal thickness of 0.6 mm or 0.024 in.) also generally failed adhesively. However, the percent adhesive remaining on side B ranged from 3 to 15%.

A notable observation concerning the thickest specimens was a number of small spherical voids, having diameters of about 0.2 to 0.5 mm (0.008 to 0.02 in.), found throughout the adhesive layers. These voids were attributed to the entrapment of the solvent during preparation of the thickest specimens. They were not observed in the adhesive layers of the thinner specimens. The thickest adhesive layers had more solvent present per unit of specimen surface area. Consistent with the estimation of bulk flow resistance using the finger test (Table 3) and mass measurements during cure time, the thickest specimens retained more adhesive solvent than the thinner specimens. Apparently, the residual solvent in the thickest specimens agglomerated before escape to form the voids. Solvents entrapped during assembly of adhesive joints [16] and also coating applications [17] are known to produce voids in adhesive or coating layers.

Scanning electron microscopy (SEM) was performed on selected rubber strips after their separation by peel testing. The SEM observations were consistent with the percent mass measurements in Fig. 4. The specimens examined and the results for the areas subjected to SEM analysis are as follows:

1. a clean specimen with a thin adhesive layer: Adhesive was observed on both rubber strips (sides A and B), consistent with the cohesive failure mode.

2. a heavily contaminated specimen with a thin adhesive layer: The surfaces of both strips A and B were covered with platelet particles, typical of the talc-like material used as the contaminant. The observation was consistent with adhesive failure occurring between the adhesive layer and the surface of strip B where the contaminant was deposited. No evidence of adhesive on strip B was found.

3. a clean specimen with a thick adhesive layer: Both rubber strips (sides A and B) were seen to be covered with adhesive, consistent with a cohesive failure mode. The voids in the adhesive were clearly visible, as is shown in Fig. 5.

4. a heavily contaminated specimen with a thick adhesive layer: The results indicated adhesive failure. The surfaces of both side A and B strips were covered with platelet particles. The voids were

clearly visible in the adhesive layer on the side A strip, as is shown in Fig. 6. The SEM analysis also showed the presence of small sections of adhesive on the side B strip (Fig. 6). This observation was attributed to failure of the adhesive layer at the edges of the voids which were close to the rubber surface. These areas of adhesive apparently were weaker than the bond strength to the contaminated surface of the rubber. Consequently, they ruptured from the bulk of the adhesive and remained on the contaminated strip. The presence of the small sections of adhesive on the side B strip would account for the increase in mass measured for the contaminated, thick specimens, as given in Fig. 4.

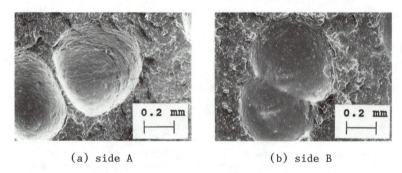

(a) side A   (b) side B

FIG. 5 -- SEM photographs of loci of failure after peel test. Nominal grayscale 30 (clean), nominal adhesive thickness 0.54 mm (0.021 in.).

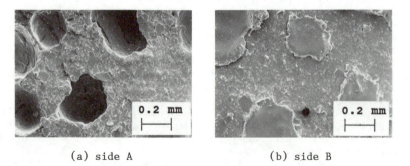

(a) side A   (b) side B

FIG. 6 -- SEM photographs of loci of failure after peel test. Nominal grayscale 160 (heavily contaminated), nominal adhesive thickness 0.54 mm (0.021 in.).

Fig. 7 is a summary of the failure modes observed for the specimens having varying degrees of surface cleanness and adhesive thickness. Consistent with previous observations [9,10], the EPDM roofing seam specimens generally failed cohesively when the surface of the rubber was clean, and adhesively when the surface of the rubber was heavily contaminated.   An additional factor from the present study is that the thickest adhesive layers contained voids which resulted in some adhesive remaining on the contaminated surface during peeling.

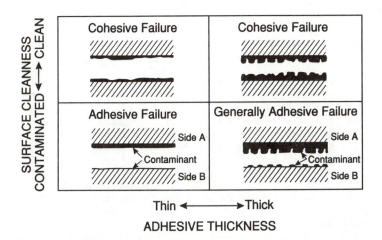

Thin ◄────────► Thick

ADHESIVE THICKNESS

FIG. 7 -- Summary of failure modes of seam specimens having various adhesive thickness and surface cleanness.

Relation between the Adhesive Thickness and Peel Strength

A review of the mechanics of peel of adhesive joints has been given by Gent and Hamed [13].   In this paper, these authors discussed the relation between adhesive thickness and peel strength. They noted with examples that "it is commonly observed that the peel force increases as the thickness of the adhesive layer is increased." The peel force becomes independent of thickness at relatively thick adhesive layers. The two authors explained this by attributing the adhesive thickness effect to energy dissipation in the bulk of the adhesive. With thicker adhesive layers, the work expended in peeling increases because a larger volume of adhesive is subjected to deformation per unit area of detachment. However, at still larger adhesive thickness, the peeling process no longer involves the entire adhesive layer, and the energy dissipated during peel becomes independent of the total adhesive thickness.

The results of the present study (Fig. 3) were consistent with the general effect of adhesive thickness as reviewed by Gent and Hamed [13].   However, in the present study, a question remained concerning the effect of the voids found in the thickest adhesive layers.   In particular, it may be asked whether the voids contributed to reducing the peel strengths of the specimens below that which may have been found in specimens without voids. Because they represent defects, it might be expected that the voids would contribute to reduced strength if they are located near the locus of failure during peeling. For example, it may be that voids concentrated in the center of an adhesive layer would have little effect in cases where the failure mode is adhesive. An investigation of the effect of the voids was beyond the scope of the present study. In a study on air entrapped voids and their prevention in some adhesive joints, Bascom and Cottington [18] reported that peel strength was greater for specimens without voids.

SUMMARY AND CONCLUSIONS

This study was conducted to provide further data on factors influencing the results of peel tests of seam specimens of EPDM rubber roofing membranes. A peel test was used, because previous studies have shown it to be sensitive to factors such as surface contamination that may be expected to affect strength. The data obtained would contribute to providing a database needed to support the development of a quality assurance methodology for seams.

The primary factors addressed in the study were thickness of the adhesive layer and open time during adhesive application. The two had received little attention in previous studies. In addition, the condition (clean and contaminated) of EPDM rubber seams was included. Contaminated specimens were prepared by depositing a talc-like material over the surface of cleaned rubber strips to simulate release agent on the membrane.

The following is a summary of the key findings and conclusions:

o No effect of open time was observed. This was particularly the case for cleaned specimens having open times ranging from 5 min to 15 h. It was cautioned that the results were obtained in a laboratory experiment to understand the factors affecting peel strength. They may not be strictly applicable to forming seams in the field where the exposed adhesive might be affected by job site conditions such as dust contamination.

o Talc-like contaminants on a surface of the rubber strips resulted in a loss of peel strength. This finding was consistent with those of previous studies. Seam specimens with cleaned rubber strips generally failed cohesively; whereas those having contaminated rubber strips generally failed adhesively.

o The peel strength of the specimens generally increased with the thickness of the adhesive layer. The relation between thickness and strength was not linear, but appeared to tend towards a limiting value, which was consistent with reported theoretical discussions.

o With regard to the combined effects of surface cleanness and adhesive thickness, the greatest peel strengths were found for clean rubber surfaces with thick adhesive layer. In contrast, the lower peel strengths occurred for heavily contaminated rubber surfaces with thin adhesive layers. Specimens which were clean and had thin adhesive layers displayed peel strengths comparable to those that were contaminated and had thick layers of adhesive. Increasing the amount of adhesive in the seam may, to a degree, overcome the adverse effect of surface contamination. This finding suggests that, if the T-peel test is used for quality control purposes, the thickness of the adhesive layer in the specimen should also be measured.

o Observations made during peel testing and confirmed using SEM analysis indicated the presence of small, spherical voids in the adhesive layers of the thickest specimens. These voids were attributed to the entrapment of the adhesive solvent in the specimens having thick adhesive layers. It was questioned whether these voids contributed to a reduction in peel strength versus that expected for specimens without the voids. However, experimentation to examine the effect of the voids was beyond the scope of the study.

ACKNOWLEDGMENTS

The authors acknowledge with thanks the assistance of their colleagues at the National Institute of Standards and Technology (NIST) who contributed to the study: Edward Embree, John Winpigler, and Jack Lee for preparing specimens; Jim Seiler for providing advice on conducting peel tests; Paul Stutzman for conducting SEM analysis; Jonathan Martin for discussing experimental plans and results, and for his review of the paper; James Lechner for conducting statistical analysis, discussing experimental results, and reviewing this paper; and Larry Masters for reviewing the paper.

REFERENCES

[1] Gish, B. D., and Jablonowski, T.L, "Weathering Tests for EPDM Rubber Sheets for Use in Roofing Applications," Proceedings, Eighth Conference on Roofing Technology, National Institute of Standards and Technology (formerly National Bureau of Standards), Gaithersburg, MD, 1987, pp. 54-68.

[2] Doherty, F. W. and Shloss, A. L., "Single-Ply Synthetic Rubber Roofing Membranes," Single-Ply Roofing Technology, ASTM STP 790, W. H. Gumpertz, Ed., American Society For Testing and Materials, Philadelphia, 1982, pp. 40-54.

[3] Westley, S. A., "Bonding Ethylene Propylene Diene Monomer Roofing Membranes: The Theory and Practice of Adhering Vulcanized Ethylene Propylene to Itself," ibid., pp. 90-108.

[4] Strong, A. G., "Factors Affecting the Jointing of Vulcanized Membranes," Proceedings, Paper No.60, Meeting of the American Chemical Society Rubber Division, Philadelphia, May, 1982.

[5] Rossiter, W. J., Jr., "Tests of Adhesive-Bonded Seams of Single-
     Ply Rubber Membranes," Roofing Research and Standards Development,
     ASTM STP 959, R. A. Critchell, Ed., American Society for Testing
     And Materials, Philadelphia, 1982, pp. 53-62.
[6] Dupuis, R. M., "Analysis and Design of Adhesive Lap Splices for
     Elastomeric Single-Ply Membranes," Proceedings, Seventh Conference
     on Roofing Technology, National Institute of Standards and
     Technology (formerly National Bureau of Standards), Gaithersburg,
     MD, 1983, pp. 16-21.
[7] Martin, J. W., Embree, E., and Bentz, D. P., "Effect of
     Temperature and Stress on the Time-to-Failure of EPDM T-Peel
     Joints," Proceedings, Eighth Conference on Roofing Technology,
     National Institute of Standards and Technology (formerly National
     Bureau of Standards), Gaithersburg, MD, April, 1987, pp. 69-74.
[8] Rossiter, W. J., Jr., "Further Investigation of the Effect of
     Application Parameters on Adhesive-Bonded Seams of Single-Ply Roof
     Membranes," Materials and Structures, 1988, Vol. 21, pp. 243-249.
[9] Martin J. W., Embree, E., Rossiter W. J., Jr., "Effect of
     Contamination Level on Strength of Butyl-Adhered EPDM Joints in
     EPDM Single-Ply Roofing Membranes," Proceedings, Ninth Conference
     on Roofing Technology, National Institute of Standards and
     Technology (formerly National Bureau of Standards), Gaithersburg,
     MD, May 1989, pp. 64-72.
[10] Rossiter, W. J., Jr., Seiler J. F. Jr., and Stutzman, P. E.,
     "Field Testing of Adhesive-Bonded Seams of Rubber Roofing
     Membranes," ibid., pp. 78-87.
[11] Rossiter, W. J., Jr., "Field Evaluation of the Quality of Newly
     Formed Adhesive-Bonded Seams of Single-ply Membranes," in Roofs
     and Roofing, May Jack O., Ed., Ellis Horwood Ltd., Chichester,
     1988, pp. 273-289.
[12] Wu, S., "Fracture of Adhesive Bond," in Polymer Interface and
     Adhesion, Marcel Dekker, Inc., New York, 1982, pp. 530-554.
[13] Gent, A. N. and Hamed, G. G., "Peel Mechanics of Adhesive Joints,"
     Polymer Engineering and Science, July 1977, Vol. 17, No. 7, pp.
     462-466.
[14] Landrock, A. H., Adhesives Technology Handbook, Noyers
     Publications, NJ, 1985, p. 208.
[15] ibid., p. 184.
[16] ibid., p. 212.
[17] Pierce, P. E. and Schoff, C. K., Coating Film Defects, Federation
     of Societies for Coatings Technology, PA, 1988, p. 18.
[18] Bascom, W. D. and Cottington, R. L., "Air Entrapment in the Use of
     Structural Adhesive Films," Journal of Adhesion, 1972, Vol. 4, p.
     193-209.

Glen D. Gaddy, Walter J. Rossiter, Jr., and Ronald K. Eby

THE APPLICATION OF THERMAL ANALYSIS TECHNIQUES TO THE
CHARACTERIZATION OF EPDM ROOFING MEMBRANE MATERIALS

REFERENCE: Gaddy, G. D., Rossiter, W. J. Jr., and Eby, R.
K., "The Application of Thermal Analysis Techniques to the
Characterization of EPDM Roofing Membrane Materials,"
Roofing Research and Standards Development: 2nd Volume, ASTM
STP 1088, Thomas J. Wallace and Walter J. Rossiter, Eds.,
American Society for Testing and Materials, Philadelphia,
1990.

ABSTRACT: This study was conducted to provide data on the
feasibility of using thermal analysis (TA) methods for the
characterization of roofing membrane materials. TA methods
have not been widely applied to such materials. The
methods used were thermogravimetry (TG), differential
scanning calorimetry (DSC), and dynamic mechanical analysis
(DMA). Three black (carbon black filled), two white
(titanium dioxide pigmented), and one white on black
laminate ethylene propylene diene terpolymer (EPDM) membrane
materials were analyzed before and after exposure to the
heat, ozone and UV conditions given in ASTM D 4637, as well
as to outdoor exposure. Load-elongation tests were
conducted to compare the results with those of the TA
methods. The results indicated that: (1) TA techniques can
be used to characterize EPDM membrane materials; (2) both
the black and white membrane materials showed only slight
property changes under exposure, as determined using the TA
methods; and (3) in contrast to the TA results, the load-
elongation values displayed relatively large changes. The
TA tests determined bulk properties of the rubber sheets,
whereas the elongation measurements were, to a great extent,
influenced by surface characteristics. Based on the study
results, it is recommended that work continue to provide
additional data necessary for use of TA methods in
standards.

KEYWORDS: ASTM D 4637, characterization, environments, EPDM
rubber, roofing, single-ply membranes, thermal analysis

Mr. Gaddy is the National Roofing Contractors Association/
National Institute of Standards and Technology (NRCA/NIST) Graduate
Student at The Johns Hopkins University, Baltimore, MD 21218; Dr.
Rossiter is a Research Chemist at NIST, Gaithersburg, MD 20899; Dr.
Eby is professor of Materials Science and Engineering at The Johns
Hopkins University, Baltimore, MD 21218.

Revolution, according to Webster [1], is a radical change. This definition accurately describes the use of single-ply membranes in lieu of multi-ply built-up roofing (BUR) as the waterproofing component in low-slope roof systems. The largest share of the single-ply market belongs to ethylene-propylene-diene-terpolymer (EPDM). According to the National Roofing Contractors Association (NRCA) [2], 45% of the low-sloped roofing applications in 1988 were EPDM. The reasons behind the increased use of EPDM are several, chief among them is its mechanical properties, such as low temperature flexibility and resistance to building movement [3]. Other advantages of EPDM include low weight, resistance to foot traffic, ease of roofing unusual contours, and the reduction in labor costs associated with simplified installation.

The increased use of new membrane materials has necessitated the development of improved characterization techniques. In the past, methods for testing elastomeric roofing membrane materials were based on ASTM Specification D 4637-87 for Vulcanized Rubber Sheet Used in Single-Ply Roof Membrane, which includes the load-elongation test used in the present study. By contrast, thermal analysis (TA) methods have not been widely utilized for the characterization of roofing membrane materials. TA is a generic term for a group of techniques which measure the change in physical properties as a function of temperature. The TA methods can provide highly reproducible results with very small sample sizes [4]. Additionally, TA techniques provide more diverse information than other methods alone, such as load-elongation or low temperature embrittlement. The TA techniques described in this paper are thermogravimetry (TG) which measures mass change, differential scanning calorimetry (DSC) which measures relative heat flux and, in turn, the heat capacity and the glass transition temperature, and dynamic mechanical analysis (DMA) which measures the moduli of the materials and, in turn, the glass transition. An additional method, thermal mechanical analysis (TMA) measures changes in volume or length as a function of temperature. This method will not be discussed further here but will be the subject of another paper.

In 1988, an international roofing committee, working under the joint auspices of CIB/RILEM[1], recommended that TA methods be investigated for use in characterizing membrane materials and the changes that may occur on aging [5]. The committee acknowledged that little research had been reported on the application of TA methods to roofing and, thus, also recommended that research was needed to provide the technical basis for this application. The present study followed, in part, from that recommendation.

Past research aimed at the application of TA methods to roofing membrane characterization includes papers by Farlling [6] and by Backenstow and Flueler [7]. The work by these authors supported the activities of the CIB/RILEM Committee. Both papers [6,7] reported

---

[1]CIB is an acronym for Conseil International du Batement pour la Recherche l'Etude et la Documentation; RILEM is an acronym for Reunion Internationale des Laboratoires d'Essais et de Recherches sur les Materiaux et les Constructions.

on the characteristics of EPDM, polyvinyl chloride (PVC), and polymer-modified bituminous materials using TG, DSC, and DMA methods. In addition, Backenstow and Flueler [7] described the application of torsion pendulum analysis to the characterization of these membrane materials. Each paper concluded that these techniques were useful for membrane characterization and should be investigated as methods for incorporation in standards.

An earlier paper on the application of TA methods to the analysis of roofing membrane materials was presented by Cash [8] at a previous ASTM D08 symposium on single-ply roofing technology. He reported preliminary DSC data for the characterization of EPDM, neoprene, chlorinated polyethylene (CPE), and polyvinyl chloride (PVC). He concluded that DSC could be used to identify the components in a single-ply sheet, differentiate between manufacturers, and differentiate between exposed and new materials.

Since it is possible to detect changes in membrane properties using TA methods, it becomes necessary to explain the significance of these changes as well as their cause and impact. It is expected that, over time, membrane material properties change. The changes may or may not be detrimental to in-service performance. One of the challenges of applying TA methods to exposed roof membrane characterization is to determine whether observed changes are acceptable.

The need for standard methods for the characterization of different types of roofing membrane materials, exposed to the same service conditions, is important. Classically, test methods have been prescriptive rather than performance oriented. As a result, many test procedures address materials by their generic type without concern for their end use. For example, the load-elongation tests for EPDM and modified bituminous materials are different, although the products often have the same application. An advantage of TA methods is that they are equally applicable to different membrane materials [6].

The present study was designed to provide further data on the application of TA methods to membrane characterization and to provide a basis for the use of the methods in performance standards for membrane materials. The application of TA methods to the characterization of roofing membrane materials is a broad subject due to many factors including the numerous materials available and application methods used. It was thus necessary to establish a study of limited scope. The study undertaken was to investigate changes which black and white EPDM membrane materials may undergo upon exposure, relative to each other, and the applicability of TA methods for following these changes. In addition, results of changes in properties determined using TA methods are compared to changes in load-elongation as classically measured for rubber sheet materials according to D 412.

EPDM is synthesized from ethylene, propylene, and a small percentage of diene, which is normally 1,4- hexadiene, ethylidene norbornene, or dicyclopentadiene [9]. The diene molecules enable the polymer to crosslink by sulfur vulcanization while maintaining a saturated polymer backbone, which provides the elastomer with

improved ozone resistance. Prior work in roofing [10] has reported
on the mechanism of attack by ozone. EPDMs are described as having
satisfactory weather and heat resistance [9], although these
properties depend upon the formulation of the EPDM rubber product.
For example, UV stabilizers must be added to formulations used for
rubber roofing sheets [11]. The addition of the carbon black adds
ultraviolet (UV) stability to black EPDM membrane materials by
providing a light screen and absorbing the UV radiation [12]. In
contrast, for white products, titanium dioxide is normally added to
provide a light screen and reflect the UV radiation [13]. Since
these UV stabilizers function by different mechanisms, their
effectiveness may depend upon compounding and the percent of
stabilizer added to the specific formulation. It has been noted
that titanium dioxide (and colored pigments) give much less UV
protection than carbon black [14].

EXPERIMENTAL METHODS

Test Samples

    Three samples of black, two samples of white, and one sample of
white on black laminate, commercially available EPDM sheet membrane
materials were obtained (Table 1). Sample 2W was a laminated-by-
vulcanization sheet of a white ply and a black ply. The samples
were provided by three manufacturers, hereafter designated as 2,3,
and 4. Upon receipt, the materials were cleaned three times by
scrubbing both sides in soap and water followed each time by a water
rinse. After drying, the samples were solvent cleaned using a
hexane-saturated cloth to remove any contaminents, especially talc,
from the surface. A total of 300 samples were cleaned and divided
equally among the various exposure conditions.

TABLE 1 -- Test Samples

Designation	Color	Thickness(mm)
2B	Black	1.5
2W	White	1.5
3B	Black	1.5
3W	White	1.5
4B	Black	1.5
4W	White	1.5

Exposure Conditions

    Laboratory exposure: Test specimens of the six membrane samples
were subjected to the heat, ozone, and UV conditions defined in ASTM
D 4637-87. Table 2 provides a summary of the exposure conditions
and includes references to the ASTM test methods specified in ASTM D
4637-87. This standard gives the length of time over which the
heat, ozone and UV exposures are to be continuously conducted in
determining conformance of a vulcanized rubber roofing sheet to the
specification. In the present study, the times of exposure were
varied (Table 3). One set of exposure times was approximately those
given in ASTM D 4637-87 (Table 2).

TABLE 2 -- Laboratory Exposure Conditions

Exposure Method	ASTM Test	Exposure Conditions Specified in D 4637
Heat	D 573	670 h at 115°C.
Ozone	D 1149	166 h at 100 mPa and 40°C at 50% elongation.
UV	G 26	2000 h at 80°C black panel; 690 min light on, and 30 min light on with water spray; 0.35 W/m² at 340 nm.

TABLE 3 -- Laboratory Exposure Times

Exposure	Times Exposed (in hours)
Heat	21, 70, 96, 168, 335, 692, & 5136.
Ozone	5, 8, 25, 49, 150, 242, & 335.
UV	67, 142, 260, 552, 1000 & 2000.

Note: Exposure times were $\pm$ 0.5 hours.

Outdoor exposure: Rectangular (25 x 150 mm) and dogbone (ASTM D 412, Die C) samples were attached to boards and mounted on an outside weathering rack at NIST. The exposure was due south at forty-five degrees from perpendicular for 6438 $\pm$ 10 h (approximately nine months, from July to March).

Thermal Analysis Methods

Thermogravimetry: TG was performed using the Perkin-Elmer[2] Model TGS-2. The specimens (5 to 15 mg) were heated from 50 to 770°C at a rate of 20°C per minute. The procedure followed the CIB/RILEM recommendation [5]. Pyrolysis was conducted in nitrogen gas at a flow rate of 40 mL/min until 600°C was reached, and then air was introduced at the same flow rate to combust the residual material. The instrument was allowed to cool to 50°C and then purged with nitrogen for about 5 min prior to the next run. Trials were performed prior to the test samples on a well characterized

---

[2]Certain trade names or company products are mentioned in the text to specify adequately the experimental procedure and equipment used. In no case does such indication imply recommendation or endorsement by The Johns Hopkins University or the National Institute of Standards and Technology, nor does it imply that the products are necessarily the best available.

EPDM and the experimental coefficient of variation was found to be $\pm$ 2 % for TG.

Differential scanning calorimetry: This analysis was performed using a Perkin-Elmer DSC-2C, which was modified to operate using a liquid nitrogen coolant. This allowed for stable temperatures down to -100°C.

Samples were prepared by using a paper hole punch to make round samples (dia. 7 mm) which were weighed and then wedged into aluminum DSC sample holders which were used without tops. The tight fit of the samples allowed for good thermal contact. Preliminary tests showed that the open top did not adversely affect the results. The specimen and holder were placed in the test chamber along with an empty pan for reference, which was washed in a flow of helium gas at a rate of 40 mL/min throughout the analysis. The specimen was stabilized at -100°C before heating at 10°C/min to a maximum temperature of 200°C. Preliminary trials were performed on well characterized membrane material and the test method was found to have experimental variability of $\pm$ 5°C.

Dynamic mechanical analysis: This analysis was conducted using a DuPont DMA 962 operated in the transverse mode from -100°C to 50°C at a heating rate of 1°C/min. The specimens were maintained at their resonance frequencies. The specimen dimensions were 1.5 x 11.5 mm. Preliminary trials were performed on well characterized membrane material and the test method was found to have experimental variability of $\pm$ 3°C.

Load-Elongation Tests

Load-elongation testing was performed according to ASTM D 412, Test Methods for Rubber Properties in Tension, (Die C) using an Instron Series IX Automated Materials Testing System. Elongation was measured using an automated extensometer with a gauge length of 25 mm. The specimens were elongated at 518 mm/min. Preliminary trials were performed on well characterized membrane material and the test method was found to have experimental variability of $\pm$ 2%.

RESULTS AND DISCUSSION

Thermal Analysis Methods

Thermogravimetry: Typical TG curves are shown in Fig. 1, which depicts specimen change in mass with increasing temperature. In this example, a thermogram for control specimen 2B is plotted along with those from the analyses of the 2B specimen after maximum exposure to the heat, ozone, UV, and outdoor conditions given in Table 2. Onset of mass loss was about 340°C which was consistent with that reported for pure EPDM rubber [15].

The curves obtained in the present study were comparable to those reported by Farlling [6] for the same heating conditions.

Specimen mass loss occurred over two distinct temperature ranges:
(1) from onset of heating to about 550°C and (2) after the addition
of air at 600°C to the termination of heating. These two ranges
divided the curve for the black samples into three compositional
groups: (1) organic materials liberated in nitrogen below 600°C, (2)
carbonaceous materials, primarily carbon black [6], readily oxidized
at temperatures above 600°C, and (3) ash which remained after
oxidation. The curves for the white materials look the same except
for the carbonaceous region which is not present. The organic
materials in the first compositional group include volatile organics
and base polymer of the roofing sheet. Variations in the percent
organic materials on exposure are important because they are
indicative of changes in the basic composition of the sheet
including the polymer structure. It is evident (Fig.1) that little
variation in the weight loss for the organic constituents was
observed for the 2B sample under any of the exposure conditions.
However, it was noted that the shape of the ozone curve differed
from the other exposure condition samples. This was due to the
increased loss of volatile organics at low temperatures. The reason
for this was not investigated; however, a reason may be that the
ozone is attacking side chain unsaturation yielding volatile
organics at lower temperatures.

Table 4 summarizes the mass percent of the three TGA
compositional groups determined for samples 2-4 (both black and
white) after maximum exposure (Table 2) as determined from curves
such as those in Fig. 1. In the case of the black controls, all
three samples had comparable organic composition (54 to 59 percent),
whereas the percent mass of the carbonaceous material (indicative of
carbon black) ranged from 26 to 39 percent.

TABLE 4 -- Mass Percent of Each Component
as Determined by TGA

Sample Design.	Exposure Condition	Organics B	Organics W	Carbonaceous B	Carbonaceous W	Ash B	Ash W
2	Control	58	58	30	15	12	27
	Heat	54	53	32	13	13	32
	Ozone	59	56	26	14	15	30
	UV	56	56	30	16	13	28
	Outdoor	56	56	31	14	13	30
3	Control	54	53	39	0	7	47
	Heat	50	50	42	0	8	50
	Ozone	53	53	39	0	7	46
	UV	52	52	39	0	8	49
	Outdoor	57	53	30	0	13	47
4	Control	59	57	26	0	15	43
	Heat	57	53	28	0	15	46
	Ozone	59	56	25	0	15	43
	UV	59	55	26	0	15	45
	Outdoor	60	57	26	0	14	43

The heading spanning: Compositional Group, Mass Percent[a,b]

[a]The B and W denote black and white, respectively
[b]See text for description of test conditions

White sample nos. 3 and 4 had similar percent compositions including no carbonaceous component and relatively high ash content (as compared to the black samples). This reflected the absence of carbon black or materials combustible above 600°C in the sheets and the presence of inorganic stabilizers such as titanium dioxide. The presence of a carbonaceous component for sample 2W was due to the presence of carbon black in the black ply laminated to the white ply of the composite membrane sheet.

Table 4 shows that the exposures of the samples caused little change in the final percent mass of the compositional groups, indicating mass stability of the sheets to the selected heat, ozone, UV, and outdoor conditions. In general, heat exposure produced the largest mass changes of organic components, which were in the range of 2 to 5 percent depending upon the sample. Note that the heat exposure was conducted for 5136 hours, which far exceeded the exposure time (670 h) required in ASTM D 4637.

Fig. 2 present plots of the mass percent of the organic component of the samples as a function of time for the heat, ozone and UV exposures to show the time of exposure dependence of mass changes. In general, consistent with the data in Table 4, no large changes were induced in the percent of organic component.

The absolute percent change of the organic component after heat, ozone, and UV exposures for the time periods given in ASTM D 4637 was determined. As shown in Table 5, the absolute changes ranged from 0 to 3 percent depending on the sample and exposure conditions. Remembering that the coefficient of variation was $\pm$ 2 %, the result compares favorably with a preliminary recommendation from the CIB/RILEM committee [5] that allowable changes (on exposure) in the percent organic component of a membrane material not exceed $\pm$2 percent. The CIB/RILEM recommendation was based on exposure conditions which were generally less severe than those specified in ASTM D 4637, and used in the present study.

TABLE 5 -- Percent Change in Organic Component as Determined by TGA after Exposure to ASTM D 4637 Conditions

Exposure Condition	Change in Organic Component, %					
	Sample 2		Sample 3		Sample 4	
	B	W	B	W	B	W
Heat	2	3	3	1	0	1
Ozone	2	3	2	1	0	0
UV	2	2	2	0	0	1

On the basis of the data in Table 5, no suggestions to ASTM Committee D08 for incorporation of a TGA test in D 4637 appear to be appropriate without more testing. To implement any suggestion, additional data may be needed from the field, especially data from materials tested before and after long exposure times. These data should indicate what material changes due to artificial exposure indicate long term performance on the roof. In this regard, black EPDM products have been available since the mid-1970s and experience has shown that the sheet materials have performed satisfactorily in service.

Differential Scanning Calorimetry: Fig. 3 shows a typical DSC thermogram, which was obtained from the analysis of an unexposed specimen of sample 4B. The curve in Fig. 3 is not corrected with a baseline correction. The key feature of the curve is the slight increase in the sample specific heat with an inflection in the curve, which ranges from about -70 to -40°C which agrees with the value reported by Maurer [15] of -51°C for pure EPDM. This inflection is indicative of the glass transition temperature (T_g) range of the material. Note that, in the present study, the T_g is taken as the midpoint of the inflection in the DSC curve. The overall shape of the curve in Figure 3 is a result of the instrument baseline instability over a wide temperature range and is not reflective of the material. Additionally, the peak occurring at approximately 30°C, which is present on sample 2B, may be due to two factors. The most likely reason for this peak is a loss of crystallinity by the semi-crystalline regions of the EPDM. This is a common event which is shown to occur over a temperature range of -40° to 40°C depending on the processing and composition of the EPDM [15]. The less likely cause would be a delayed melting peak for condensed water on the surface of the sample. This is especially unlikely since the result was only seen for sample 2B. The glass transition is associated with the polymer changing from a glassy to a rubbery state [16]. Below the glass transition, the polymer has properties typical of glasses including hardness, stiffness, and brittleness. As synthetic polymers undergo aging in service, they may increase in brittleness and related properties. A means of following such changes is to measure the T_g using a technique such as DSC. From a practical standpoint, substantial increases in T_g of a roofing membrane material may be detrimental to long-term performance if the materials are brittle at normally encountered service temperatures.

Table 6 presents T_gs (as measured by DSC) for the six unexposed EPDM samples, along with the maximum changes in T_g due to heat, ozone, UV, and outdoor conditions. The positive values in Table 6 indicate an increase in T_g. Changes in T_g ranged from -2 to 12 °C depending on the material and the exposure conditions. This range of changes is not considered significant in predicting membrane performance for the following reasons. First, in the case of the unexposed EPDM roofing sheets, the T_gs are relatively low so that a small increase in brittleness may not adversely affect in-service performance. Second, in discussing allowable changes in the T_g of roofing sheets subjected to laboratory exposure, the CIB/RILEM Committee [17] considered that variations in T_g of ± 10 °C might be acceptable. This suggestion was based on preliminary data obtained from the analysis of products considered to provide satisfactory performance. The values in Table 6 are consistent with the original CIB/RILEM discussions.

TABLE 6 -- Change in Glass Transition Temperature
on Maximum Exposure, as Measured by DSC

Sample Design.	Control Tg, °C	Change in Glass Transition, °C			
		Exposure Condition			
		Heat	Ozone	UV	Outdoor
2B	-64	5	8	10	9
2W	-59	12	10	9	12
3B	-66	4	10	3	3
3W	-57	7	--[a]	0	0
4B	-60	4	-2	6	--
4W	-71	12	6	2	--

[a]The dashed lines indicate no data available.

An important question is the change in T_g as a function of
exposure time. Data (not reported herein) from the DSC analysis
indicated that little change occurred during exposure, similar to
the results found for TGA. In this regard, Table 7 presents changes
in T_g experienced by the samples after heat, ozone, and UV exposure
to the time periods given in ASTM D 4637. Note the similarity of
these data with those in Table 6; the greatest difference was 10 °C
for sample 2W exposed to ozone. Note also in Table 7 that the
maximum change in T_g due to exposure to the ASTM conditions was 10
°C, which is the limit proposed by CIB/RILEM [17]. This suggests to
ASTM Committee D08 that a DSC requirement for incorporation in D
4637 consider a maximum allowable change in T_g of ± 10 °C. However,
the DSC measurement of T_g is redundant with other TA methods such as
TMA and DMA. Moreover, it was found, in the present study, that
the DSC method was less practicable, for example, requiring longer
set-up and testing times, than the other methods such as TMA.
Therefore, it is recommended that a method such as TMA be considered
to determine T_g and not the DSC method.[3]

TABLE 7 -- Change in Glass Transition Temperature on
Exposure to ASTM D 4637 Conditions, as Measured by DSC

Sample Design.	Change in Glass Transition, °C		
	Exposure Condition		
	Heat	Ozone	UV
2B	6	6	10
2W	10	7	9
3B	4	10	3
3W	8	6	0
4B	4	-2	6
4W	10	6	2

[3]Paper in Preparation

Dynamic Mechanical Analysis: Analysis using this technique was
limited. Samples 3W and 4B were examined before exposure (controls)
and after exposure to the heat, ozone, and UV conditions and time
periods prescribed in ASTM D 4637. Typical results of a DMA
analysis are shown in Fig. 4 for the control sample 4B. The
thermogram in Fig. 4a shows two curves: resonance frequency of the
specimen and damping. Damping is a measure of the energy absorbed
by a material. As the specimen undergoes a glass transition, a peak
in energy absorption occurs [18]. The glass transition is, thus,
represented by the peak in the damping curve.

For specimen 4B, the glass transition temperature occurred at
about -43°C (Fig 4a). Note that this value is 17°C higher than that
measured by the DSC method (Table 6). This was not unexpected. The
measure of glass transition depends upon the method [19]. It has
been recognized that DMA provides a T_g that is somewhat higher than
that obtained by DSC [6].

The thermogram in Fig. 4b shows three curves: storage modulus
(E'), loss modulus (E''), and tan delta. E' is proportional to the
peak stored energy, E'' is proportional to the net energy dissipated
per cycle of deformation, and tan delta is the ratio of E''/E' [19].
A key feature of the thermogram is the peak in the tan delta curve.
It is a measure of T_g, analogous to the peak in the damping curve.

Table 8 presents a summary of the T_g values for samples 3W and
4B as measured by the peaks in the damping and tan delta curves.
The two methods provided comparable values, which was expected
because they were derived from the same measurement. For purposes
of standards development, only one would be necessary. The changes
in the T_g due to heat, ozone, and UV exposure did not exceed 7°C,
which was within the limits of 10°C proposed by CIB/RILEM [5].

TABLE 8   -- Glass Transition Temperatures Measured as the
    Peak of the Damping Curve and the tan delta Curve

Sample Design.	Exposure Conditions	T_g(°C) Damping	tan delta
3W	Control	-43	-41
	Heat	-39	-38
	Ozone	-40	-41
	UV	-38	-39
4B	Control	-43	-39
	Heat	-36	-35
	Ozone	-44	-39
	UV	-40	-38

As an indication of changes induced in a membrane material due
to exposure, storage modulus curves for exposed specimens may be
qualitatively compared to that of the control [18]. The comparison
was made below T_g where the modulus values were considerably greater
than zero. The results of the comparison are summarized in Table 9
for the control and exposed samples 3W and 4B. The table indicates
whether the storage modulus shifted to higher or lower values in

relation to that of the control. The relative changes in modulus found upon heat and ozone exposure were expected due to their known hardening and softening effects, respectively, on the material [6, 21]. It was expected that UV exposure would also result in a higher modulus, due to a crosslinking of the polymer [21]. As is evident in Table 8, a lowering of the storage modulus was observed upon UV exposure. The reasons for this observation are not known and further investigation was beyond the scope of the study. If DMA is to be used for incorporating a modulus requirement into a consensus standard, the reasons should be understood.

TABLE 9 -- Change in Storage Modulus Upon Exposure

Sample Design.	Exposure Condition	Change in Modulus Relative to the Control
3W	Heat	Higher
	Ozone	Lower
	UV	Lower
4B	Heat	Higher
	Ozone	Lower
	UV	Lower

Load-Elongation Tests

Load-elongation tests were conducted to allow comparison of the results of a classical method with those obtained by thermal analysis. In the case of EPDM roofing membrane materials, elongation is often the parameter used as an indicator of change occurring during exposure. For example, ASTM D 4637 requires that the elongation be determined after heat exposure. It has been shown from field data that maximum elongation of EPDM membrane materials is more affected during exposure than is tensile strength [22,23].

Table 10 presents the results of the ultimate elongation measurements. As evident from the table, the control samples displayed a wide range of elongation, from 570 to 980 percent. Data are not available (Table 10) from all samples under all exposure conditions because of experimental difficulties encountered during the study.

Table 10 -- Ultimate Elongation (%) for Materials Subjected to Maximum Exposure

Sample Design.	Control	Elongation, % Exposure Conditions			
		Heat	Ozone	UV	Outdoor
2B	790	100	610	520	780
2W	600	--	600	480	400
3B	570	80	550	310	550
3W	980	--	900	--	560
4B	620	--	--	410	540
4W	850	210	810	630	730

All exposure conditions resulted in a decrease in elongation. The largest decrease (about 80 percent) was observed for the heat exposure (which was conducted for 5136 h). Strong and Puse [22] have reported that any loss in properties for EPDM membrane materials is most likely due to heat from either temperature exposure or solar radiation. They have shown data whereby specimens heated at 116 °C for 2352 h, had elongations of less than 100 percent.

Ozone generally had little effect on elongation (Table 10), consistent with the known resistance of EPDM membrane materials to ozone [9]. The outdoor exposure of the samples gave variable results, with elongation values ranging from essentially no change to about a 40 percent decrease. Rosenfield [23] has presented data showing that black EPDM samples exposed outdoors for 4320 h experienced losses in elongation of about 10 percent of the original value. It is noted that, in the present study, the greatest decreases were found for 2 of the white samples (2W & 3W).

The results of the TG and DSC tests (Tables 5 and 7, respectively) may be compared with the elongation measurements (Table 10). As previously discussed, the percent organic components and the T_g values showed little change on exposure, regardless of the conditions. In contrast, the elongation values displayed relatively large changes, depending on the exposure condition, as just mentioned above. A possible reason for the relative differences in the magnitude of the property change (TA vs elongation measurements) is believed to associated with the properties which are measured in the two types of tests. The TA tests in question determine bulk properties of the rubber sheet, whereas the elongation measurement is, to a greater extent, influenced by the its surface characteristics. In the latter case, changes to the surface such as embrittlement during exposure may result in the significant decreases in maximum elongation. The results from the study, thus, suggest that TA methods might be more appropriate for examining whether changes in bulk properties are induced during exposure.

Tables 11 and 12 give ultimate stress and modulus values for the control and exposed samples. Consistent with previous data [21,22,23] on the laboratory and outdoor exposure of EPDM roofing membrane materials, the ultimate stress and modulus values decreased and increased, respectively.

Table 11-- Ultimate Stress (MPa) for
Materials Subjected to Maximum Exposure

Sample Design.	Control	Ultimate Stress, MPa Exposure Conditions			
		Heat	Ozone	UV	Outdoor
2B	12.7	7.8	10.1	12.4	12.2
2W	10.8	--a	9.3	9.9	6.8
3B	11.9	10.2	10.2	11.2	12.3
3W	11.9	--	11.0	--	6.5
4B	9.9	--	--	10.1	9.4
4W	13.4	10.5	12.7	9.8	11.8

aDashed lines indicate no data.

Table 12 -- Modulus at 300% Elongation (MPa) for
Materials Subjected to Maximum Exposure

Sample Design.	Control	Modulus, MPa Exposure Conditions			
		Heat	Ozone	UV	Outdoor
2B	1.7	--[a]	2.1	2.8	1.8
2W	2.0	--	1.8	2.5	1.9
3B	2.3	--	2.3	3.9	2.5
3W	1.3	--	1.4	--[b]	1.9
4B	1.9	--	--[b]	3.1	2.1
4W	1.8	--	1.8	2.2	2.0

[a]Values not reported since the heat-exposed
specimens never reached 300% elongation.
[b]Dashed lines indicate no data.

SUMMARY AND CONCLUSIONS

This study was conducted to provide data on the feasibility of
using TA techniques for the characterization of roofing membrane
materials. Black and white EPDM membrane materials were selected
for the study. They were chosen because they comprise the largest
segment of the single-ply membrane market, and the two colors of
EPDM sheet represent different means of UV stabilization. Three
test samples of each color were subjected to heat, ozone and UV
conditions given in ASTM D 4637, and also to outdoor exposure.

TA methods have not been widely applied to roofing membrane
materials. The TA methods used in the present study were TGA, DSC,
and DMA. A question raised was whether the black and white EPDM
materials exhibited differences in response to the exposure
conditions, as detected using the TA methods. In addition, load-
elongation tests were conducted to compare the results with those of
the TA methods. Based on the results of the study, the following
conclusions may be drawn:

o Both the black and white membrane materials showed only slight
  change under the exposure conditions, as determined using the TA
  methods. The changes in percent organic component and glass
  transition temperature were generally within the limits
  suggested by the CIB/RILEM Committee. No significant difference
  was found between black and white membrane materials exposed in
  this study.

o In contrast to the percent organic components and the Tg values
  determined using TA methods, the percent elongation values
  displayed relatively large changes. These TA tests determined
  bulk properties of the rubber sheet, whereas the elongation
  measurement was, to a greater extent, influenced by the surface
  characteristics. The results from the study indicated that the
  TA methods were more appropriate for determining changes in bulk
  properties induced during exposure than the elongation
  measurement.

For the integration of TA into ASTM standards for EPDM roofing membrane materials, considerable work is necessary to determine the property changes which are and are not acceptable. This will be possible only with the analysis of samples before and after aging. Without this data, it is impossible to predict the performance of these materials over time.

ACKNOWLEDGEMENTS

The authors to thank the National Roofing Contractors Association, the National Institute of Standards and Technology and The Johns Hopkins University for supporting this study. Of special note at the NRCA are William C. Cullen, Robert LaCosse, and The Technical Operations Committee which have enthusiastically supported the work. At NIST, the work was assisted by the efforts of W. Eric Byrd, S.S. Chang, Willard E. Roberts, and James Seiler. Additionally at NIST, the comments of Mary McKnight, Tinh Nguyen, and Larry Masters were very valuable and greatly appreciated. Special thanks are extended to James Lechner for his many valuable discussions of the experimental plan and data analysis. The DMA measurements were provided by Mohammed Ibrihim and Monica Klemens at Martin Marietta Laboratories, Baltimore, Maryland.

REFERENCES

[1] Webster's New Collegiate Dictionary, Merriam Co., Springfield, MA, 1961.

[2] Cullen, W.C., "Project Pinpoint Analysis: Trends and Problems in Low Slope Roofing 1983-88," NRCA, Rosemont, IL, 1989.

[3] Kenney, H., "Ethylene Propylene Diene Terpolymer," Roofing Spec, May 1981.

[4] Wendlandt, W.W. and Gallagher, P.K., "Instrumentation," in Thermal Characterization of Polymeric Materials, E.A. Turi, Ed., Academic Press, New York, 1981, pp. 1-86.

[5] "Performance Testing of Roofing Membrane Materials," Recommendations of CIB W.83 and RILEM 75-SLR Joint Committee on Elastomeric, Thermoplastic, and Modified Bitumen Roofing, RILEM, Paris, France, November 1988.

[6] Farlling, M. S., "New Laboratory Procedures to Evaluate the Durability of Roofing Membranes," Appendix D in "Performance Testing of Roofing Membrane Materials," Recommendations of CIB W.83 and RILEM 75-SLR Joint Committee on Elastomeric, Thermoplastic, and Modified Bitumen Roofing, RILEM, Paris, France, November 1988.

[7] Backenstow, D. and Flueler, P., "Thermal Analysis for Characterization," Proceedings, 9th Conference on Roofing Technology, National Roofing Contractors Association, Rosemont, IL, May 1989, pp. 85-90.

[8] Cash, C.G., "Thermal Evaluation of One-Ply Sheet Roofing," Single-Ply Roofing Technology, ASTM STP 790, W.H. Gumpertz, Ed., American Society for Testing and Materials, Philadelphia, PA, 1982, pp. 55-64.

[9] Babbitt, R.O., Ed., The Vanderbilt Rubber Handbook, R.T. Vanderbilt, Norwalk, CT, 1978, p. 536.

[10]Laaly, H.O., "Methods of Evaluating Single-ply Roofing Membranes," Single-Ply Roofing Technology, ASTM STP 790, W.H. Gumpertz, Ed., American Society for Testing and Materials, Philadelphia, PA, 1982, pp. 65-75.

[11]Gish, B.D. and Jablonowski, T.L. "Weathering Tests for EPDM Rubber Sheets for Use in Roofing Applications," Proceedings, 8th Conference on Roofing Technology, National Roofing Contractors Association, Rosemont, IL, April 1987, pp. 54-68.

[12]Hawkins, W. L., "Environmental Deterioration of Polymers" in Polymer Stabilization, W.L. Hawkins, Ed., Wiley-Interscience, New York, 1972, pp. 1-28.

[13]Clark, H.B., "Titanium Dioxide Pigments" in Treatise in Coatings, Vol. 3, Pigments, R.R. Myers and J.S. Long, Eds., Marcel Dekker, New York, 1975, pp. 480-531.

[14]Baseden, G.A., "Compounding NORDEL Hydrocarbon Rubber for Good Weathering Resistance", Du Pont, Wilmington, DE, 1987.

[15]Maurer, J.J., "Elastomers," in Thermal Characterization of Polymeric Materials, E.A. Turi, Ed., Academic Press, New York, 1981, pp. 572-705.

[16]Billmeyer, F.W., Textbook of Polymer Science, 2nd Ed., Wiley, New York, 1971, pp. 8-10.

[17]Minutes of the 28-29 April 1988 Meeting, CIB/RILEM Committee on Elastomeric, Thermoplastic and Polymer Modified Bituminous Roofing, RILEM, Paris, France, April 1988.

[18]Read, B.E. and Dean, P.D., The Determination of Dynamic Properties of Polymers and Composites, Wiley, New York, 1978, pp. 17-19.

[19]Hill, L.W., "Mechanical Properties of Coatings," Federation Series On Coating Technology, Federation of Societies for Coating Technology, Philadelphia, PA., 1987, p. 17.

[20]Meyer, J. et al., "Advanced Mechanical Analysis Techniques for the Viscoelastic Characterization of Polymers," American Laboratory Vol 22,No 1, January, 1990, pp. 60-63.

[21]Doherty, F.W. and Shloss, A.L., "Single-ply Synthetic Rubber Roofing Membranes," Single-Ply Roofing Technology, ASTM STP 790, W.H. Gumpertz, Ed., American Society for Testing and Materials, Philadelphia, PA, 1981, pp. 40-54.

[22]Strong, A.G. and Puse, J.W., "Outdoor Exposure of EPDM Roofing Membrane," Proceedings, Second International Symposium on Roofing Technology, National Roofing Contractors Association, Rosemont, IL, 1985, pp. 376-382.

[23]Rosenfield, M.J., "Field Test Results of Experimental EPDM and PUF Roofing," Proceedings, Second International Symposium on Roofing Technology, National Roofing Contractors Association, Rosemont, IL, 1985, pp. 275-279.

Lyle D. Hogan

EVALUATING POLYMERIC ROOFING MEMBRANES FOR LINEAR DIMENSIONAL STABILITY

REFERENCE: Hogan, Lyle D., "Evaluating Polymeric Roofing Membranes For Linear Dimensional Stability", Roofing Research and Standards Development, 2nd. Volume, ASTM STP 1088, Thomas J. Wallace and Walter J. Rossiter, Eds., American Society for Testing and Materials, Philadelphia, 1990.

ABSTRACT: Currently available polymeric roofing membranes are ordinarily evaluated in a battery of tests that are hopefully indicators of in-situ performance. Field performance of some products has suggested that dimensional stability may be important among these. The test for such a property (ASTM D-1204) is a hot air bath environment for measured membrane samples. Presently, test environments are selected by manufacturers and range widely in the market. To promote congruency in testing/reporting, a scale is proposed for future testing within categories defined by generic polymer type.

KEY WORDS: Thermoset, Thermoplastic, Uncured Elastomer, Co-Polymer Alloy.

Specifiers or purchasers of currently available polymeric roofing membranes may expect confusion when evaluating the products. Though responsible selection involves consideration of a family of such properties, dimensional stability is of particular importance for many synthetic membranes. As indicated in the Significance and Use Statement of the Standard, the test is an indicator of

Mr. Hogan is a Certified Roof Consultant and Senior Engineer with Trigon Engineering Consultants, Inc., Post Office Box 18846, Greensboro, North Carolina 27419.

lot-to-lot uniformity regarding internal strains introduced during processing; however, the perception of the property (by some members of the specifying community) is the shrinkage tendency of a product. A representative mode for shrinkage failure is shown on photos 1 through 6.

The simple test involves membrane samples in a hot air bath with final dimensions compared to original dimensions on a percentage change basis. The test title specifically cites "Thermoplastic Sheeting", but thermosetting polymers and uncured elastomers also recognize and use the test.

Thermoplastic materials are those which may be repeatedly heated (softened) and cooled (hardened) without irreversibly changing physical properties. These materials include PVC, and the family of PVC blends such as EIP (Ethylene Interpolymers) and NBP (Acrylonitrile Butadiene Polyblend) which contain PVC to varying amount [1]. These are tabulated in Chart B as CPA's (co-polymer alloys). For purposes of this study, the relatively newer heat- weldable EPDM must be included by definition.

Thermosets could experience irreversible physical change by temperature cycling stated earlier. These are vulcanized materials representing several products available, including conventional EPDM used in this study.

Uncured elastomers are those which are initially heat or solvent weldable, yet may be vulcanized in either factory or field settings. This generic group of membranes is represented by CSPE, PIB, and CPE.

Sample Designation	Published Dimensional Stability	Environment
636	0.3%	14 days @ 168°F
642	0.2%	14 days @ 168°F
668	0.5%	24 hrs. @ 158°F
671	0.5%	166 hrs. @ 158°F
680	1.0%	6 hrs. @ 176°F
689	<0.5%	6 hrs. @ 80°C (176°F)
694	0.0%	(Not Stated)
699	<0.5%	72 hrs. @ 176°F
705	1.0%	1 hr. @ 212°F
713	0.1%	24 hrs. @ 129°F
718	2.0%	14 days @ 240°F
723	0.1%	28 days @ 212°F

Chart A

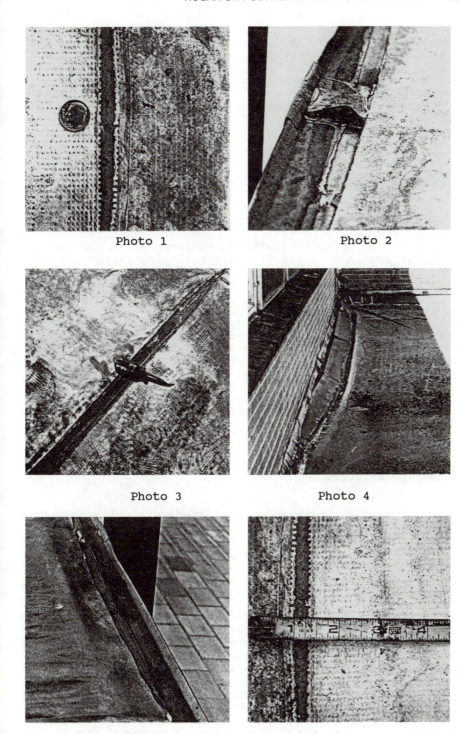

Photo 1

Photo 2

Photo 3

Photo 4

Photo 5

Photo 6

Popular figures for reporting dimensional stability are normally less than 1% (perhaps in an effort to maintain confidence of the specifier). Yet further inquiry will show a wide spectrum of testing environments used to arrive at the figure. Selected membranes (itemized on chart A above) show test environments ranging from 1 hour @ 212°F to 28 days @ 212°F. Clearly a membrane exposed to the latter has endured a more brutal environment than that realized by the former.

OBJECTIVE OF RESEARCH

Shrinkage of bituminous membranes is well documented in several sources [2], [3], [4]. Not so well documented is shrinkage in one-ply polymeric membranes which the author has observed in a number of instances. Yet reported values for dimensional stability for membranes involved fell within the popular reporting range referenced earlier.

The purpose of this study is to produce comparative data for the materials selected by simultaneous testing in an arbitrary, uniform environment. A discussion of our selected test environment is given in a later section.

If it serves no other purpose, this paper will bring to light wide variations in testing environments utilized in single ply roofing.

MATERIALS SELECTED

The 21 membrane samples evaluated were collected from specifier's manuals, trade shows, and construction projects. Thus the longitudinal and transverse directions could not always be identified. Averaging the pair of values for eventual reporting reduces the importance of knowing the orientation for purposes of this work.

Our testing originally focused on thermoplastic polymers such as the growing sector of co-polymer alloys (blended PVC's) and uncured elastomers, the last of which have been associated with a greater propensity to shrink [5] than competing membranes. Also of interest was the behavior of a relatively new family of heat-welded EPDM products, thought to be made thermoplastic by co-dispersion with polypropylene prior to co-extrusion. It is noteworthy here that these products publish the most brutal test environment of those materials evaluated herein. These are listed in Chart A as samples 718 and 723.

Although unreinforced thermosetting polymers capable of elongating in excess of 300 percent do not tend to fail in a shrinkage mode, conventional thermoset EPDM was included in this test as a comparison instrument.   As stated earlier, this product was the sole representative of the thermoset group.

PROCEDURES

A long-standing argument questions the validity of certain laboratory tests as indicators of in-service behavior. This has been true of accelerated aging by concentrated, reflected sunlight (EMMAQUA Test) [6] and highly concentrated heat and ultra-violet radiation in QUV testing apparatus.   In the absence of suitable performance track records, such arbitrary procedures were a necessary implementation.

Since reflective membranes experience thermal gain (in good draining installations) only marginally in excess of ambient temperature, a testing temperature in excess of 150°F did not (in the author's view) represent a plausible environment. A temperature of 135°F was adopted.

Samples were conditioned in compliance with ASTM D-1204.   Critical dimensions were obtained and the test chamber (see photos 7, 8) was placed into a "BLUE MAX STABIL-THERM CONSTANT TEMPERATURE CABINET".   Standard laboratory atmosphere conditions were in compliance with the test conditions required. Testing conditions were monitored 1 to 3 times daily over a period of 14 days. Samples were again conditioned prior to collecting final measurements.

Photo 7   Photo 8

RESULTS

Chart   B and the graphed results offer some   noteworthy comments:

-- Reinforced   polyethylene-based products tended toward   more dimensional change   than did   reinforced
products derived from, or based upon, PVC.   This is
exemplified in samples 625, 626,   and 627   compared
to samples 642 - 699.

-- Unsupported   products   experienced   greater   dimensional change than their reinforced counterparts.

-- Testing   variance can be significant within a   particular vendor's   product as exemplified by samples
625,   626, 627.   This group represents a 30% variation based upon the greatest value.

-- Dimensional   loss   in one direction can   be   offset
partially   or completely by dimensional gain in another   direction.   This is exemplified   in   paired
figures having opposite signs.   In one case,   a net
gain in measurements was recorded.

-- The heat weldable EPDM samples,   along with the co-
polymer   alloys resulted in comparatively low   values.

-- Uncured   elastomers represented both   greatest   and
least values observed.   This is evident by samples
730 and 733.

-- Products having fleece backing or fibrous glass reinforcements exhibited lowest values recorded.

-- Products   having weft-inserted   reinforcing   scrims
resulted   in   lower values than   those   with   woven
scrims.

CONCLUSIONS AND RECOMMENDATIONS

Taken as a whole, the figures are conservative and less
than   initially expected.   Based on experience   gained   in
this study,   the selected temperature of 135°F was too mild
to   conclusively   isolate products given   to   comparatively
higher dimensional change.   Further comprehensive study is

necessary to evaluate more appropriate test climates.   It could follow that the test climate suitable for identifying dimensional change in uncured elastomers is inappropriate for inducing similar behavior in pure thermoplastics.

The issue of correlation to field performance is prone to arise with future standardization efforts.   Therefore the test duration variable should also be explored.   Consensus from the manufacturing sector would enhance the acceptability of any stages of testing environment such as the scale below:

GENERIC MEMBRANE TYPE	TESTING ENVIRONMENT
Pure Thermoplastics (including co-polymer alloys)	$135^{\circ} - 145^{\circ}$ F
Uncured Elastomers	$165^{\circ} - 175^{\circ}$ F
Thermosets, possibly including heat-weldable EPDM blends	$195^{\circ} - 205^{\circ}$ F

Further testing is also needed to evaluate performance variation due to reinforcement types.   It appears that categorizing a product by this feature may be as indicative of field behavior as polymer formulation.

Since most roofing membranes experience eventual shrinkage to some extent, responsible specifiers should have uniform testing results from which to evaluate susceptibility for dimensional change. While relief of pent-up internal strains is the goal of the test, the figures are reported in sales literature and subject to interpretation by the specifier or buyer.   The current flexibility to change test environment may obscure unimpressive properties that this test might otherwise identify.

REFERENCES

[1] Russo, Michael, "Categorizing the Newer Generation of PVC Blends", RSI, October 1987, p. 64.

[2] R.G. Turenne, "Shrinkage Of Bituminous Roofing Membranes". Canadian Building Digests (Paper No. 181).

[3] NRCA Project Pinpoint (Problem Job Report Summary).

[4] W.H. Gumpertz, "Performance Considerations For Thermally Efficient Roofing Systems". Proceedings Of The Fifth Conference On Roofing Technology (Paper No. 7), April 1979, Page 49.

[5] Myer J. Rosenfield, "Initial Investigation Of 3 Uncured Elastomer Roofing Membrane Materials For Use In Military Construction". U.S. Army Corps of Engineers Interim Report M86/03 February 1986, pp. 12, 16.

[6] EMMAQUA Test Method ASTM E-838-81.

Sample Designation	Generic Polymer Type	Chemical Composition	Reinforcement Type	Dimensional Change x-y (%)	a-b (%)	Average (%)
625	Uncured Elastomer	CPE	Weft-Inserted Polyester	-0.43	-1.22	-0.82
626	Uncured Elastomer	CPE	Weft-Inserted Polyester	-0.61	-0.90	-0.75
627	Uncured Elastomer	CPE	Weft-Inserted Polyester	-0.13	-1.02	-0.57
642	Thermoplastic	CPA	Weft-Inserted Polyester	-0.15	-0.12	-0.13
668	Thermoplastic	CPA	Weft-Inserted Polyester	-0.14	-0.12	-0.13
671	Thermoplastic	CPA	Weft-Inserted Polyester	0.00	-0.24	-0.12
680	Thermoplastic	CPA	Weft-Inserted Polyester	0.00	-0.12	-0.06
694	Thermoplastic	PVC	Fibrous Glass	-0.26	+0.24	-0.01
699	Thermoplastic	CPA	Woven Polyester	-0.54	+0.90	+0.18
713	Uncured Elastomer	CSPE	Woven Polyester	+0.27	-0.60	-0.16
718	Thermoplastic	EPDM/P.P.	None	0.00	-0.12	-0.06
723	Thermoplastic	EPDM/P.P.	None	0.00	-0.12	-0.06
729	Uncured Elastomer	PIB	None	-0.13	+0.11	-0.01
730	Uncured Elastomer	PIB	Polyester Fleece Backing	0.00	0.00	0.00
731	Uncured Elastomer	CSPE	Woven Polyester	-0.56	-0.13	-0.34
733	Uncured Elastomer	CSPE	None	+2.11	-6.12	-2.01
735	Thermoset	EPDM	None	-0.13	-0.19	-0.16
737	Uncured Elastomer	CSPE	Woven Polyester	-0.67	0.00	-0.33
739	Thermoplastic	PVC	None	+0.40	-1.43	-0.51
740	Thermoplastic	PVC	Fibrous Glass	0.00	0.00	0.00
742	Thermoplastic	EPDM/P.P.	None	-0.13	0.00	-0.07

CHART B

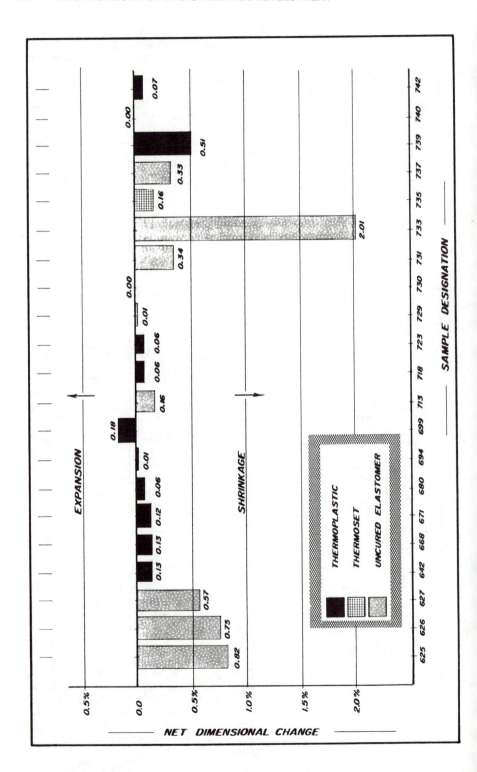

Stuart Foltz and David M. Bailey

POLYVINYL CHLORIDE (PVC) ROOFING: PRELIMINARY FIELD TEST RESULTS

REFERENCE: Foltz, S., and Bailey, D.M., "Polyvinyl Chloride (PVC) Roofing: Preliminary Field Test Results," Roofing Research and Standards Development: Second Volume, ASTM STP 1088, Thomas J. Wallace and Walter J. Rossiter, Eds., American Society for Testing and Materials, Philadelphia, 1990.

ABSTRACT: Polyvinyl chloride (PVC) single-ply roofing systems have been installed at three U.S. military facilities for a ten year evaluation. This paper reports preliminary results through the first six years of testing. The test is part of long-term research by the U.S. Army Construction Engineering Research Laboratory (USA CERL) to identify alternatives to conventional built-up roofing for military construction and reroofing.

Membrane samples were taken periodically based on a schedule of every six months for the first two years and annually during the next four years. Laboratory testing was performed by the U. S. Bureau of Reclamation using American Society for Testing and Materials (ASTM) methods in most cases. A total of ten tests measuring mechanical and physical properties of the membranes and the seams were performed on each set of samples. Four other tests were done on a limited basis.

KEYWORDS: polyvinyl chloride (PVC) membranes, single-ply roofing, flat roofing, roofing system

INTRODUCTION

The use of nonconventional membranes as alternatives to built-up roofing (BUR) increased dramatically in the 1970's and 1980's. Estimates show single plys and modified bitumen materials have from 40% to 65% of the total low slope roofing market. This growth has occurred despite improvements in BUR materials and construction specifications. Recognizing this trend, the U.S Army Construction Engineering Research Laboratory began investigating roofing materials in the late 1970's.

One of the roofing materials selected for long term field testing was polyvinyl chloride (PVC) membranes. In 1982 and 1983, test roofs were installed on buildings in three different areas of the country: Chanute AFB, IL, Dugway Proving Ground, UT, and Fort Polk, LA. Three manufacturers were chosen and two products from each manufacturer were used. Laboratory testing was performed on

Mr. Foltz and Mr. Bailey are researchers at the Construction Engineering Research Laboratory of the U.S. Army Corps of Engineers, P.O. Box 4005, Champaign, IL 61821.

membrane samples taken periodically from each test roof for six
years, measuring mechanical and physical properties and their changes
over time. This report discusses the results from this testing of
PVC roof systems.

DESCRIPTION OF TEST ROOFS

The three different sites for the test program were selected for
geographic and climatic diversity. A project building was selected at
each of the three sites. Each building roof was divided into three
sections allowing membrane material from each of the three
manufacturers (A, B, & C) to be installed. The roof sections ranged
in area from 6,932 to 13,642 square feet (644 to 1,267 square
meters).

The roofing systems at Chanute AFB consist of a poured-in-place
concrete deck, a two ply organic felt and asphalt vapor retarder, and
2-1/2 inch (6.35 cm) of aluminum foil faced isocyanurate foam board
in two layers, loose laid. The PVC membranes were installed loose
laid and ballasted. All three manufacturers required the use of
membranes specifically formulated for ballasted systems. For system
A, the membrane manufacturer required no slip sheet, however the
membrane from manufacturer B required a reinforced kraft paper slip
sheet between the membrane and insulation. In lieu of using a slip
sheet for system C, the top layer of insulation was faced with 3/4
inch (1.9 cm) perlite board, installed with the perlite up. See
figure 1.

At Dugway Proving Ground, the systems consist of a poured-in-
place concrete deck, 3 inch (7.6 cm) of aluminum foil-faced
isocyanurate foam board in two layers mechanically fastened to the
decks without a vapor retarder, and the membrane. Systems A and B
were mechanically fastened. The system A membrane was adhered to 7
inch (17.8 cm) diameter discs of reinforced PVC membrane anchored by
the insulation fasteners. The system B membrane which had a kraft
paper slip sheet, was welded to mechanically attached steel battens.
The system C membrane was fully adhered. See figure 2.

The systems at Fort Polk consist of a tongue-and-grove wood plank
deck and 4-1/2 inch (11.4 cm) of aluminum foil-faced expanded
polystyrene insulation board in two layers which are mechanically
fastened to the decks without a vapor retarder. Specific membrane
materials were the same as those at Dugway Proving Ground. The
system A membrane, which had a fiberglass slip sheet under it, was
mechanically fastened along one edge of each sheet, with the next
sheet lapped over the washers and welded to the first sheet. System
B required a reinforced kraft paper slip sheet and was attached using
the batten system similar to that at Dugway. For system C, the top
layer of insulation was faced with 1/2 inch (1.3 cm) of fiberboard,
installed with the fiberboard up (requiring no slip sheet). The
membrane was fully adhered to the fiberboard surface. Systems B and
C were installed on an arched roof. Because of the curvature, the
layers of insulation were fastened independently. (See figure 3.)

Table 1 gives a simplified description of the roof systems. A
more complete discussion of the construction of the test roofs can be
found in USACERL Technical Report M-343, April 1984, "Construction of
Experimental Polyvinyl Chloride (PVC) Roofing" [1] and further
information can be found in Technical report M-87/04, "Experimental
Polyvinyl Chloride (PVC) Roofing: Field Test Results" [2].

DESCRIPTION OF MANUFACTURED PVC MEMBRANES

In 1986, ASTM published specification D 4434 [2] that categorizes
PVC membranes into three types, I, II, and III. Type II is
subdivided into two grades.

Type I: Unreinforced sheet
Type II:
          Grade 1- Unreinforced sheet containing fibers
          Grade 2- Unreinforced sheet containing fabrics
Type III: Reinforced sheet containing fibers or fabric

Contrary to the above quoted ASTM specification, type II membranes
are often called reinforced membranes. Membranes used in this field
test include all three types. Each type was used at all three
locations except that type III was not used at Chanute. See table 1.
Most manufacturers specify an unreinforced membrane for ballasted
systems.

Although an individual manufacturer often varies the formulation,
having materials from multiple manufacturers assures that the
membranes will have different amounts of various plasticizers and
processing oils (from here on referred to as only plasticizers) and a
different degree of polymerization. Differences in manufacturer
formulation and the addition of reinforcements categorized by the
three ASTM types previously listed can result in materials with
substantially different characteristics. It can therefore be said
that PVC membranes are not a single product and more properly
describe a family of products. The unfortunate result of this is
that conclusions from the testing cannot be immediately assumed to
apply to all PVC membranes and must be scrutinized carefully to
evaluate their applicability to other PVC membranes. Other variables
besides membrane formulation which will likely affect the results
include the roofing system, climate, quality of construction, and the
type and use of the building.

DESCRIPTION OF TEST PROGRAM

The test program was designed to determine how weathering would
change the mechanical and physical characteristics of the various PVC
membranes. An initial set of tests was performed on each of the
different materials soon after installation to establish material
characteristics of new PVC membranes. Samples were taken from each
different material at each site on a periodic basis: every six months
for the first two years and annually for the next four years. Five
one square foot samples were taken from each roof section. One
sample was taken from near each corner and the fifth was cut at the
center of the roof.

Laboratory testing was performed by the U. S. Bureau of
Reclamation using American Society for Testing and Materials (ASTM)
methods. A total of 14 tests measuring mechanical and physical
properties of the membranes and the seams were performed on each set
of samples. See table 2.

The properties which were selected for study were considered to
have the potential to indicate the performance level of the membrane
materials of a roof assembly. The dimensional stability and
plasticizer loss tests were performed initially but were not done
again until year three. The tensile, elongation, and tear
resistance tests were done only on type I and II membranes and the
ply adhesion test was only used for the type III membrane.

GENERAL PVC PROPERTY CHANGES

Data resulting from the testing is presented on graphs which show
the linear best fit lines for the data. Individual data points
followed the linear fit very well and were not shown on the graphs to
improve clarity. Data for the plasticizer loss and dimensional
stability tests was limited and considered to be more informative in
table form.

Most of the properties tested had results which generally followed the expected trends. The tensile strength (Fig. 4,5,6) and tear resistance (Fig. 7,8,9) had increases of 5 to 35 percent with only one exception (Manufacturer B at Dugway). Membrane B at Dugway had a very small increase in tear resistance and the tensile strength actually dropped slightly. The ballasted systems at Chanute showed a greater tendency to increase in tensile strength and tear resistance as compared to the exposed membranes at Dugway and Ft. Polk. The unreinforced membranes of manufacturer A (at Chanute) and B (all sites) had larger decreases in elongation. In general, the elongation (Fig 10,11,12) decreased 10 to 15 percent for the type I unreinforced membranes and 5 to 10 percent for the type II membranes (manufacturer C).

Although the same type III reinforced membrane by manufacturer A was used at both Dugway and Ft. Polk, testing of the ply adhesion gave substantially different results (Fig 13). Until the third year, the samples from Ft. Polk fell short of the 10 lb./in. (1.75 kN/m) width claimed by the manufacturer. After 36 months, it had increased 25 percent to 10 lb./in. At Dugway, the ply adhesion decreased over time but was still well above the claimed value.

Changes in abrasion loss (Fig 14,15,16) varied greatly between membranes over the six years of testing but no membrane showed a pattern of decreased abrasion loss. Contrary to the earlier statement on the closeness of fit for all data, abrasion loss data for these membranes had numerous wide variations from one year to the next and large standard deviations. It is not known if this problem is due to variability of the samples, laboratory procedures used for these tests, or inherent problems with the ASTM test procedure for abrasion loss. Despite suspicions raised by the large variability in the data, it can be observed with relative confidence that all of the membranes used in this field test had increased abrasion loss over time. Although large or increasing abrasion loss is undesirable, it is not known if these values are significantly large enough to result in inadequate performance. Increased abrasion loss could be attributable to the membrane changes caused by plasticizer loss.

Seam shear strength for all membranes (Fig 17,18,19) increased from 3 to 20 percent except for material A at Ft. Polk which increased almost 40 percent and material B at Dugway which decreased approximately 15 percent. At initial testing, seam shear strength was above the recommended 80 percent of the membrane tensile strength except for the membrane of manufacturer A at Chanute which was barely over 70 percent. This is not a significant weakness but does raise the question of the quality and completeness of the seam. This shear strength increased in later tests.

The seam peel strength (Fig 20,21,22) decreased 10 to 25 percent for the solvent welded membranes (membranes by manufacturer A and B at Ft. Polk and by manufacturer B at Dugway) over the test period. Seam peel strengths of the remaining membranes (all heat welded) decreased 20 to 60 percent except for B at Chanute which increased 65 percent. This increase should not be presumed significant because only three sample sets were taken from this membrane and only one set deviated from the trend of decreasing strength. Comparing the lost strength in pounds per linear inch shows a much more pronounced difference in the two techniques. The solvent welds lost from 1.8 to 3.4 lbs./in. width (0.31 to 0.6 kN/m) and the heat welded lost from 9 to 31 lbs./in. (1.58 to 5.43 kN/m). The heat welded seams were initially stronger but have consistently lost strength. The solvent welded seams were initially weaker, but have been relatively unchanged. For the heat welded membranes, the progressive decrease from the initial strength through year six is a major concern. This property will be investigated closely at site visits and in future testing. Obviously, one possible cause of any seam problems is workmanship, but the seaming method (heat versus solvent) could also be a factor in long term performance.

The membrane samples taken from the test roofs were not particularly conducive to the peel test. This test is often more accurately called a T-peel test. Normally, the vertical portion of the T is two attached strips of membrane and the top of the "T" is unattached parts of those two strips which are pulled. However, when samples from a roof are used, the seam overlap results in unattached parts of the strips at opposite ends of the welded vertical portion of the "T". There are three techniques which can be used to successfully test the samples. If the overlap made during construction was not fully welded toward the inside of the roof during construction or there was extra overlap, there may be enough free membrane so that the grips can hold the sample. Sometimes a weak part of the seam can be partly separated by hand. If this fails, freezing the material makes the weld easier to separate. The last resort is to weld on an extension piece. During this test program, samples requiring the extension piece could only be successfully tested about one half the time. Rarely did this result in less than three samples of the five from any one roof system being tested.

A significant change in the membranes of manufacturers A and B was the ineffectiveness of solvent welding the aged membrane. (Solvent welding of seams of material C was not approved by the manufacturer.) This was observed in the repairs made to the test cuts. As the membranes aged, heat welding was necessary for adequate results. The ineffectiveness of solvent welding may be due to plasticizer loss. It is at minimum a restriction on how a membrane can be repaired but may also be an indicator of the degree of change in some properties of the membrane which are undesirable such as increased hardness and brittleness.

Both the plasticizer loss and dimensional stability tests (Table 3) are greatly limited by the inability to measure actual and cumulative changes of the membrane on a roof. In addition, the test conditions cannot even be correlated to in-place use. Any of this information would be more useful than what is learned from the tests. The laboratory tests only give information on relative changes occurring under laboratory conditions. These tests were not done at 6, 12, 18, and 24 months so trends are not apparent.

A substantial reduction in plasticizer loss occurred at some point after the initial tests. This is attributable to the absence of most of the easily removable plasticizers due to previous losses. Plasticizer loss at Chanute in years 4 and 5 reduced to an average of 33 percent of the initial value. At Ft. Polk, the reduction in years 3, 4, and 5 is only to 68 percent of the initial value. This indicates a definite difference in the formulation and/or in service conditions for ballasted systems versus systems with the PVC membrane exposed. It is expected that this difference will not have a pronounced affect on the long term performance but will continue to be observed.

Due to the inability of the plasticizer test to measure the actual and cumulative loss of the membrane on the roof, it was decided in mid-1989 to run a plasticizer content test. Results are in table 4. This test was run using membrane which was still on original rolls and the roof membrane samples taken in the sixth year. Original rolls of membrane were available for Chanute and Ft. Polk but not for Dugway. Because the same membranes were used at both Ft. Polk and Dugway, the original plasticizer content for Ft. Polk was also assumed for Dugway. Due to the possibility of differences in production dates of the materials at Ft. Polk and Dugway, changes in plasticizer content at Dugway have a larger degree of uncertainty than desirable.

The plasticizer content tests confirm that the ballasted systems lost more plasticizer than the exposed PVC. They also show losses of plasticizers greater than that indicated by the plasticizer loss test. This is preliminary evidence that the plasticizer loss test is not as severe as conditions on a roof. Material C had relatively small changes in the plasticizer loss tests compared to materials A and B. Results of the plasticizer content test show this is not due to smaller losses of plasticizers in this material. This indicates another possible problem with the plasticizer loss test and its applicability to conditions on the roof.

The dimensional stability test is a measurement of the percent change in length of a sample that is heated for a specified time. The data does show that the unreinforced membranes (material A at Chanute and material B at all three sites) continue to have the potential to change significantly in dimension after over four years. Type II and III membranes show little or no dimensional change during laboratory testing.

Most manufacturer specifications for the installation of type I PVC membranes include unrolling the membrane sheets and allowing them to relax for various time periods of at least one half hour. To reduce shrinkage, Army Corps of Engineers Guide Specifications (CEGS) 07555 [3] also requires that PVC membranes be unrolled and allowed to relax for at least 1/2 hour before attachment. On site inspection reports verify that 80 foot sections shrank 7 to 12 inches (17.8 to 30.5 cm). Our laboratory testing shows the possibility of substantial further shrinkage in membranes over 4 years old. A second problem with the unreinforced membranes was wrinkling. The manufacturing process of calendaring or extruding and rolling excessively stressed the membrane and left permanent deformations which caused the membrane to be nonuniform after shrinkage. This greatly increased seaming difficulties. The automatic seaming machines did not work well on the type I membranes and this is at least partly due to the nonuniformity.

Vapor transmission showed a decrease with age for all the membranes (Fig. 23,24,25). Membranes at Chanute all had decreases greater than 40 percent. At Dugway and Ft. Polk, all membranes had decreases of 20 to 25 percent except for the membrane by manufacturer A at Ft. Polk which decreased 15 percent. Vapor transmission changes may be an indication of actual loss of plasticizers and the degree to which the in-place membranes have been affected by this loss.

The nominal thicknesses of all membranes ranged from 45 to 50 mils (1.1 to 1.3 mm). Measurements on the initial samples (Fig. 26,27,28) ranged from 45 to 56 mils (1.1 to 1.4 mm). The thickness measured after six years was generally between 2 and 5 mils (4 to 10 percent) less than initially found for the new membrane. Although the changes were small, there was a definite reduction in thickness. The one exception was membrane C at Ft. Polk, it remained virtually unchanged. It is reasoned that the thickness reduction could have a direct relationship with the loss of plasticizers and/or abrasion loss.

The increase in the measured specific gravity (Fig. 29,30,31) can also be attributed to the loss of the more volatile plasticizers. The membrane by manufacturer C at Ft. Polk had no change in specific gravity. It also had no change in thickness and half as much change in elongation as other membranes. The steadiness at Ft. Polk is likely attributable to obtaining a relatively stable level of plasticizers in the membrane before initial testing was done three

months after construction. Although data is limited, the membrane by
manufacturer C at Ft. Polk had a much smaller plasticizer loss in the
initial tests and it was somewhat smaller in later years than for the
other membranes. (See table 3.) This correlation between changes in
thickness, specific gravity, elongation, and plasticizer loss
increases confidence in the accuracy but these results are also
surprising because the exact same membrane type was used at Dugway
and it had changes similar to other membranes for these tests.

There is a definite relationship between membranes used in
ballasted versus unballasted systems and the degree of change in
material properties. The changes in ballasted membranes were
generally larger. This is true for tensile strength, tear
resistance, specific gravity, thickness, vapor transmission, and
plasticizer loss. There was no apparent and consistent difference in
the changes for elongation, abrasion loss, or dimensional stability.
For abrasion loss there was a large variability in data and the
dimensional stability data is limited, making these two properties
difficult to evaluate. For elongation, there is only one membrane
(manufacturer B at Dugway) which does not follow the trend of larger
changes for ballasted systems. That membrane is also the same one
which had a decrease in the tensile strength. It is likely that
there is some abnormality with this membrane and elongation is more
affected by ballast than UV sunlight.

The loss of plasticizers results in changes in other membrane
properties. All membrane properties (not including seam strength and
ply adhesion) increased or decreased as expected except for the few
noted variations. The degree to which these changes would occur and
the effect on performance were uncertain. These changes do not yet
seem to be a factor for concern after six years.

VISUAL INSPECTIONS

Each test roof was inspected by personnel from USACERL annually.
It was learned from the site visits that the aged membrane could not
be adequately solvent welded. At Chanute and Dugway, they used heat
welding for all repairs. At Ft. Polk, they used the solvent welding
method for patching. Patches made after four years exposure at Ft.
Polk were very difficult to adhere and starting in the sixth year,
all patching was done using heat welding. Heat welded patches are
performing well.

An important PVC related finding was that puncture resistance was
limited. Some membranes appeared very susceptible to punctures,
having some punctures and numerous indentations. Any membrane which
has puncture resistance as poor as some of these PVCs would make a
poor choice under conditions of heavy foot traffic and all PVCs
should be scrutinized carefully for puncture resistance. Future
research on the performance of PVCs should include investigation of
this property.

After one to two years of exposure at Ft. Polk, the membranes had
darkened in color due to fungal growth. The membrane of manufacturer
B was most affected and was heavily blackened. This growth is in
part due to the wet subtropical climate but the strong adhesive
property is attributed to the concentration of plasticizers on the
membrane surface. Fungal growth gradually reduced as the
plasticizers gradually evaporated and/or washed away and further
plasticizer loss reduced. After five years, the fungal growth was
gone and the membrane color had returned to a more normal light grey.
This experienced led to including the ASTM fungus resistance test in
CEGS 07555.

COLD WEATHER PERFORMANCE

In December 1989, much of the country was under the effects of a record cold spell. The Midwest was hit hard. One result of these low temperatures was the catastrophic failure of some PVC membranes. Preliminary information indicates that these membranes were unreinforced ballasted PVC membranes of 40 mil thickness or less and were six or more years old.

To the extent known by the authors, all of the catastrophic failures in December 1989 and in previous years, occurred on unreinforced ballasted PVC's. Since most ballasted PVC's are unreinforced and likewise most unreinforced PVC's are ballasted, it is difficult to determine the affect each factor has on failure. Gerhard Pastuska [5] concluded that a mixture of water, mud, micro-organisms, and oxygen increased plasticizer loss and can lead to failure. Logically, unreinforced membranes would be more likely to have catastrophic failures without reinforcement to arrest splits or strengthen cracks. There is evidence to support the conclusion that both factors are involved in cold weather catastrophic failures.

CERL's PVC field test was not unaffected by the low temperatures. Membrane A at Chanute (ballasted, unreinforced) suffered splitting failure over nearly 100 percent of the roof section. This failure was not like the shattering of glass into very small pieces as had been reported on a failure in a previous year but was divided into larger pieces. Although it could be simply a coincidence, it is interesting to note that the failed membrane had the lowest percentage of plasticizer content of any membrane in the field test based on samples taken six months prior to the failure.

CONCLUSIONS

After six years, the performance of the PVC membranes used in this test had been very good until the recent failure of the one membrane at Chanute. The membranes in Louisiana, Utah, and Illinois were subjected to diverse climates, varying building usage, and different deck and insulation types. All of these variables appear to have had little effect on water tight performance except the one catastrophic failure.

REFERENCES

[1] Rosenfield, Myer J., An Evaluation of Polyvinyl Chloride (PVC) Roofing, Technical Report M-343/ADA145406 (USA CERL, 1984).

[2] Rosenfield, M. J., and J. Wilcoski, Experimental Polyvinyl Chloride (PVC) Roofing: Field Test Results, Technical Report M-87/04 (USA CERL, 1987).

[3] ASTM D 4434, Standard Specification for Poly(vinyl Chloride) Sheet Roofing," ASTM Annual Book of Standards (American Society for Testing and Materials [ASTM], 1985.

[4] Polyvinyl Chloride (PVC) Roofing, Corps of Engineers Guide Specification (CEGS)-07555 (Dept of the Army [DA], Office of the Chief of Engineers [OCE], Jan. 1989).

[5] Pastuska, Gerhard, Roof Coverings Made of PVC Sheetings: The Effect of Plasticizers on Lifetime and Service Performance, Proceedings of the Second International Symposium on Roofing Technology, 1985.

Table 1

Roofing Systems

Membrane Manuf.	Test Site	Deck	Insulation	Vapor Retarder	Membrane Attachment	Membrane Reinforcement	Seam Welds
A	Chanute	Concrete	Isocyanurate Slip sheet/fbgls	2-ply/organic	Ballasted	Type I	Heat
B	Chanute	Concrete	Isocyanurate Slip sheet/kraft	2-ply/organic	Ballasted	Type I	Heat
C	Chanute	Concrete	Isocyanurate Perlite board	2-ply/organic	Ballasted	Type II	Heat
A	Dugway	Concrete	Isocyanurate	None	Mech. attach/ lap covered	Type III	Heat
B	Dugway	Concrete	Isocyanurate Slip sheet/kraft	None	Batten strips	Type I	Solvent
C	Dugway	Concrete	Isocyanurate	None	Fully adhered	Type II	Heat
A	Polk	Wood	Polystyrene Slip sheet/fbgls	None	Mech attach/ disks	Type III	Solvent
B	Polk	Wood	Polystyrene Slip sheet/kraft	None	Mech. attach/ batten strips	Type I	Solvent
C	Polk	Wood	Polystyrene Fiberboard	None	Fully adhered	Type II	Heat

Table 2 -- Testing Specifications

Tensile Strength	ASTM D 882
Tear Resistance	ASTM D 1004
Elongation	ASTM D 882
Ply Adhesion	ASTM D 413
Abrasion Loss	ASTM D 3389
Seam Strength (Shear)	ASTM D 882
Seam Strength (Peel)	ASTM D 1876
Plasticizer Loss	ASTM D 1203
Plasticizer Content	ASTM D 3421
Dimensional Stability	ASTM D 1204
Water Vapor Transmission	ASTM E 96
Water Absorption	ASTM D 570
Thickness	ASTM D 1593
	ASTM D 751
Specific Gravity	ASTM D 792,A-1

Table 3 -- Plasticizer Loss and Dimensional Stability

	AGE (Months)		
Membranes at Chanute AFB	0	58	68
Plasticizer Loss (% weight)			
Manuf. A	8.1	... 2.1	2.84
Manuf. B	7.7	... 2.1	2.45
Manuf. C	4.3	... 1.4	1.7
Dimensional Stability (%)			
(longitudinal)			
Manuf. A	-2.4	... -3.2	...
Manuf. B	-2.4	... -0.35	...
Manuf. C	-0.1	... -0.3	...
(transverse)			
Manuf. A	+0.9	... -0.5	...
Manuf. B	+0.9	... -1.7	...
Manuf. C	0.0	... +0.1	...

	AGE (Months)			
Membranes at Dugway	0	38	52	62
Plasticizer Loss (% weight)				
Manuf. A	...	3.5	2.5	3.36
Manuf. B	...	3.2	2.7	3.18
Manuf. C	...	1.7	1.5	1.51
Dimensional Stability (%)				
(longitudinal)				
Manuf. A	-0.4	-0.2	0.0	...
Manuf. B	-1.8	-0.8	-2.2	...
Manuf. C	-0.3	-0.1	0.0	...
(transverse)				
Manuf. A	0.0	0.0	0.0	...
Manuf. B	+0.8	+0.1	-0.2	...
Manuf. C	+0.6	0.0	0.0	...

Table 3 -- continued

Membranes at Ft. Polk	AGE (Months)			
	3	36	48	60
Plasticizer Loss (% weight)				
Manuf. A	4.9	3.5	2.9	3.82
Manuf. B	4.4	2.8	2.0	2.59
Manuf. C	1.8	1.4	0.9	1.29
Dimensional Stability (%)				
(longitudinal)				
Manuf. A	-0.4	-0.2	0.0	...
Manuf. B	-2.1	-1.4	-1.2	...
Manuf. C	-0.1	-0.1	0.0	...
(transverse)				
Manuf. A	0.0	-0.2	0.0	...
Manuf. B	+1.0	-0.6	-0.4	...
Manuf. C	+0.1	-0.1	+0.5	...

Table 4 -- PLASTICIZER CONTENT (Weight Percent)

	Manuf. A	Manuf. B	Manuf. C
Chanute			
Original	31.3	32.2	34.6
68 month	22.7	25.7	25.3
Ft. Polk			
Original	29.8	30.7	33.6
60 month	25.1	26.1	29.0
Dugway			
62 month	26.3	28.4	31.0

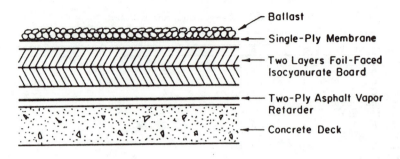

Ballast
Single-Ply Membrane
Two Layers Foil-Faced Isocyanurate Board
Two-Ply Asphalt Vapor Retarder
Concrete Deck

System A

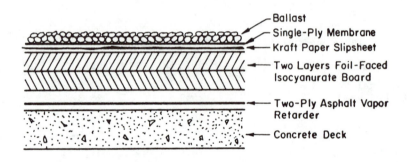

Ballast
Single-Ply Membrane
Kraft Paper Slipsheet
Two Layers Foil-Faced Isocyanurate Board
Two-Ply Asphalt Vapor Retarder
Concrete Deck

System B

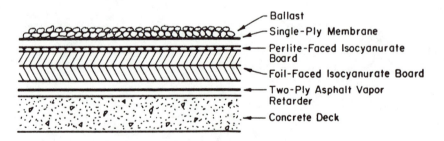

Ballast
Single-Ply Membrane
Perlite-Faced Isocyanurate Board
Foil-Faced Isocyanurate Board
Two-Ply Asphalt Vapor Retarder
Concrete Deck

System C

Figure 1. Roofing systems at Chanute AFB.

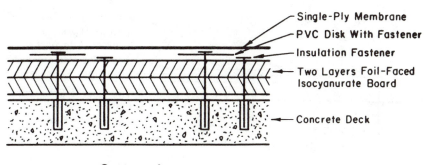

Single-Ply Membrane
PVC Disk With Fastener
Insulation Fastener
Two Layers Foil-Faced Isocyanurate Board
Concrete Deck

System A

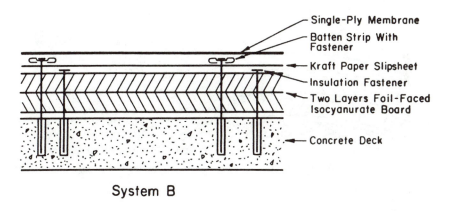

Single-Ply Membrane
Batten Strip With Fastener
Kraft Paper Slipsheet
Insulation Fastener
Two Layers Foil-Faced Isocyanurate Board
Concrete Deck

System B

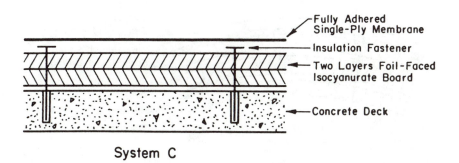

Fully Adhered Single-Ply Membrane
Insulation Fastener
Two Layers Foil-Faced Isocyanurate Board
Concrete Deck

System C

Figure 2. Roofing systems at Dugway Proving Ground.

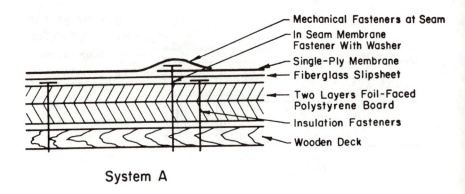

System A

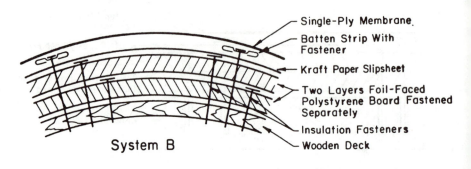

System B

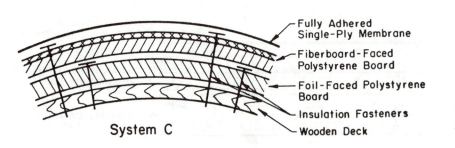

System C

Figure 3. Roofing systems at Fort Polk.

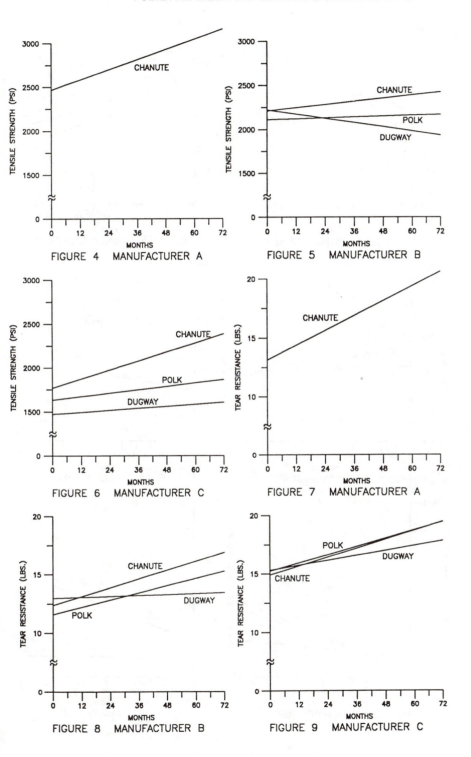

FIGURE 4   MANUFACTURER A

FIGURE 5   MANUFACTURER B

FIGURE 6   MANUFACTURER C

FIGURE 7   MANUFACTURER A

FIGURE 8   MANUFACTURER B

FIGURE 9   MANUFACTURER C

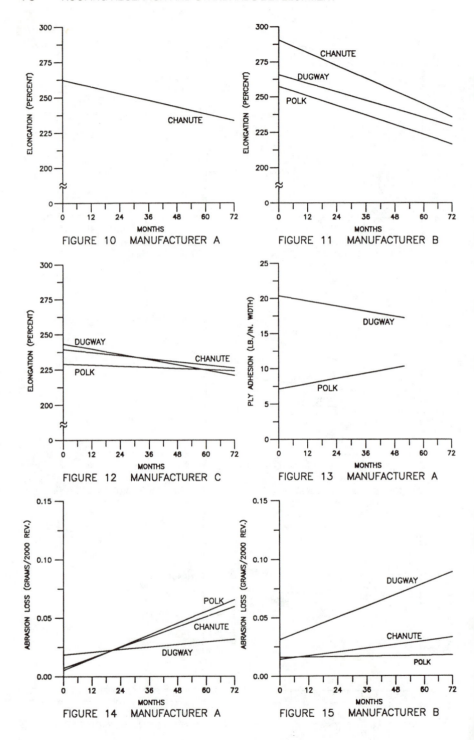

FIGURE 10   MANUFACTURER A

FIGURE 11   MANUFACTURER B

FIGURE 12   MANUFACTURER C

FIGURE 13   MANUFACTURER A

FIGURE 14   MANUFACTURER A

FIGURE 15   MANUFACTURER B

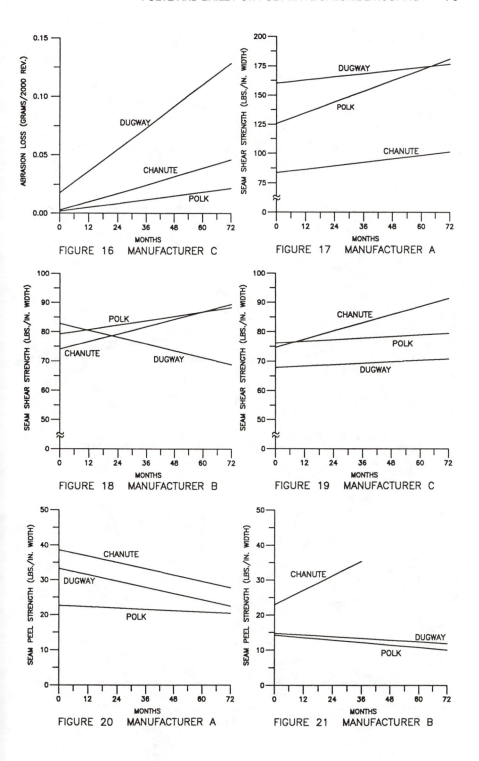

FIGURE 16   MANUFACTURER C

FIGURE 17   MANUFACTURER A

FIGURE 18   MANUFACTURER B

FIGURE 19   MANUFACTURER C

FIGURE 20   MANUFACTURER A

FIGURE 21   MANUFACTURER B

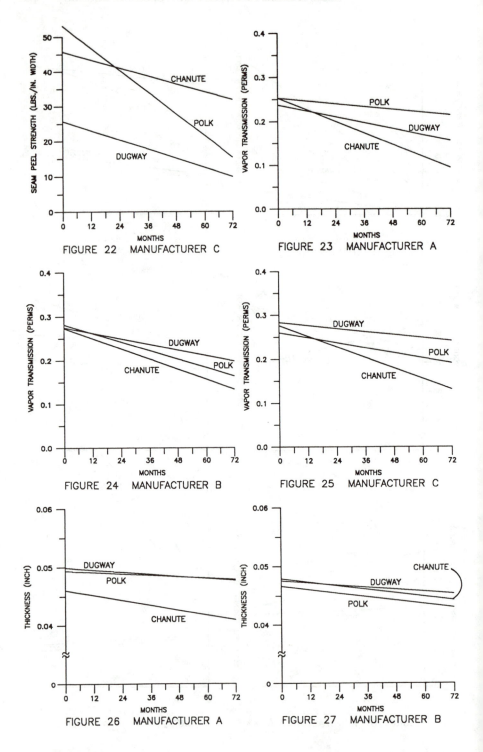

FIGURE 22   MANUFACTURER C

FIGURE 23   MANUFACTURER A

FIGURE 24   MANUFACTURER B

FIGURE 25   MANUFACTURER C

FIGURE 26   MANUFACTURER A

FIGURE 27   MANUFACTURER B

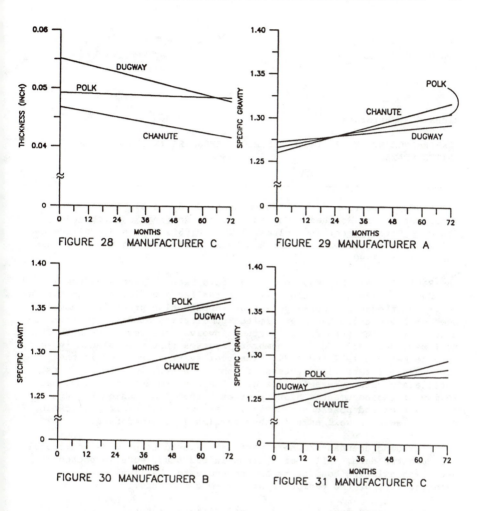

FIGURE 28   MANUFACTURER C

FIGURE 29   MANUFACTURER A

FIGURE 30   MANUFACTURER B

FIGURE 31   MANUFACTURER C

Equivalent metric quantities for measurements with English units in figures 4 to 31 are:

Fig. 4 to 6	2000 psi	= 13.8 MPa
Fig. 7 to 9	15 lbs.	= 66.7 N
Fig. 13	10 lbs./in width	= 1.75 kN/m
Fig. 17 to 19	80 lbs./in width	= 14.0 kN/m
Fig. 20 to 22	10 lbs./in width	= 1.75 kN/m
Fig. 26 to 28	0.05 in.	= 1.3 mm

Richard H. Peterson

PERFORMANCE STUDY OF SINGLE PLY ROOF MEMBRANES IN THE NORTHEAST
UNITED STATES

REFERENCE: Peterson, R.H., "Performance Study of Single Ply Roof
Membranes in the Northeast United States, "Roofing Research and
Standard Development: 2nd Volume, ASTM STP1088, Thomas J. Wallace and
Walter J. Rossiler, Eds., American Society for Testing and Materials,
Philadelphia 1990.

ABSTRACT: Single Ply roof membranes have been in extensive use in
the Northeast Region of the United States for a period of over twelve
years. Although numerous systems and application procedures have
been employed there has been limited publication of field performance
data on in-situ systems. This study involves the review of both EPDM
roof membranes and PVC roof membrane systems that were placed in
service between 1977 and 1984. The roof systems included in the
study are all reroofing systems of known construction, contractors,
materials manufacturers and design. The performance study included a
review of the documented history of each roof from design through
construction and post construction. Warranty claims, if applicable,
were reviewed as well as all other available recorded data.

    Building occupants/Owners were interviewed to provide additional
undocumented data. All roof systems in the study were inspected to
establish existing conditions and to evaluate the level of
deterioration of the various roof system components.

    Accumulated data from documented history, interviews and field
observations will be analyzed in this paper and defects will be
summarized. Causation of defects and deterioration were also
considered and classified into one or more of the catagories of
design defficiency, material performance, improper workmanship and/or
lack of appropriate and regular maintenance.

    Conclusions relative to the performance of the systems studied
will be presented and considerations given to the effect on
performance by known system changes.

KEYWORDS: Single ply, Ethylene Propolene Diene Monomer (EPDM),
Polyvinyl Chloride (PVC), Elastomeric, Plastomeric, Insulation,
Asphalt, Warranty, Useful Life, Roof, Reroof, Performance.

    R.H. Peterson is a Registered Professional Engineer specializing
in Building Envelope Systems at Gale Associates, 8 School
Street, P.O. Box 21, Weymouth, MA 02189.

INTRODUCTION

This single ply roof membrane performance study has been
performed on a controlled set of roof systems placed in service in
the Northeastern United States between 1977 and 1984. A total of 36
roof systems have been included in this study which includes review
of the original design documents, project historical data, including
construction records and warranty claims and a review of the current
conditions of the in-situ systems.

The EPDM roof membranes included in this study include both
fully adhered systems had loose laid, ballasted systems. A total of
five (5) different membrane manufacturers systems were included in
the study. The PVC roof membrane systems included in the study, are
all fully adhered systems and all were products of one manufacturer.

All of the roof systems included in the performance study were
designed by one consulting engineering firm. The limitation of
studying only those roofs that were designed by one consulting firm
allowed the establishment of a study control as all other factors
produced variables.

All of the systems included in this study are reroof systems and
all but three included removal of the existing roofing, flashing and
insulation systems prior to construction of the new system. All
three roofs that were left in place and overlaid were built-up
bituminous membranes installed in moppings of asphalt directly to
cast in place structural concrete roof decks.

The study group included buildings of varied ages, types, sizes
and uses. Structural systems included masonry bearing walls, wood
frame, steel frame, and concrete (cast in place and precast) with
total roof areas ranging from 500m^2 to 12,000m^2. Roof elevations
also varied from 4m to 50m above grade and had varying types of wind
exposure, and varying slopes from less than 0.02m/m to as much as
0.5m/m.

Roof drainage systems were typically composed of interior
through roof drain assemblies leading to a drain leader pipe system.
Other drainage types included perimeter scuppers, gutters and
downspouts and perimeter run-off.

All of the systems have been installed over board type
insulation. Insulation systems included polyurethane foam board with
facers (facers included asphalt saturated organic felts, asphalt
coated glass fiber felts, non-saturated glass fiber felt, aluminum
foil, and coated kraft paper), urethane/perlite composite board,
polyisocyanurate foam board with glass fiber felt facer and
fiberboard both regular density and high density. Insulation systems
have been installed in single and multiple layer fashion, loose laid,
mechanically attached and set in asphalt. See Table 3 for a summary
of insulation types used with the various roof systems. The total of
insulation types exceeds the number of roofs as in some roofs there
was more than one insulation system utilized due to building
conditions, deck types and fire rating requirements.

DOCUMENT REVIEW AND INTERVIEW SUMMARY

System evaluation was initiated by compiling available documents
including project specifications, contractor submittals, warranty
and/or guarantee forms, correspondence relative to system performance
and claims against the warranty/guarantee issued by the roof membrane
manufacturer.

Of specific concern during review of the available documents was
to determine the general conformance between system as designed, as
submitted for review prior to installation and as ultimately
installed. The primary focus of the document review ultimately was
to determine if the system had experienced problems necessitating a
warranty claim reaction by the warranty issuer to that claim and to
correlate with the data gathered during field condition evaluation of
the system.

The document review revealed that nine (9) of the EPDM roof
systems had been evaluated during or immediately subsequent to
construction for apparent system defects including seam delamination,
poor adhesion and delamination of flashing membranes from
substrates. Included in the systems evaluated for deficiencies
during or immediately subsequent to construction was a fully adhered
EPDM system where the field membrane was delaminating due to a
failure of the insulation skin to properly bond to the
polyisocyanurate foam insulation. Another EPDM system exhibited a
failure of the partially cured neoprene flashing sheets to adhere to
pressure treated wood curbs. The curbs were new and laden with
moisture, for exceeding the allowable level of 19% by dry weight,
from the treatment process and it was determined that it was
evaporation/ migration of this moisture which resulted in the
disbonding of the flashing sheet from the curbs.

The remaining seven defective systems were observed between the summer of 1983 and the spring of 1984. The problems observed were first noted during construction with what appeared visually to be weakness in the seaming of the field membrane. The field seams were sampled and analyzed both qualitatively and quantitatively by the membrane manufacturer with no observable deficiency in the cleaning of the seam areas prior to application of splicing cement, they reported sufficient quanity and uniform distribution of splicing cement. The samples were also analyzed for shear and peel resistance and all samples reportedly exceeded the minimum requirements of the manufacturer.

Although no reason for the seam problems was immediately apparent there was a continuation of noted seam related defects which typically initiated at three way splices in the membrane system and/or at the intersection of field seams with factory prefabricated splices. A small opening would develop in the splice at these locations and then would be infiltrated by moisture thus effecting the integrity of the neoprene based splicing cement and resulting in total delamination of the seam.

On four of the seven roofs with extensive seam problems all of the cured sheet field seams were stripped in utilizing 0.3m strips of partially cured neoprene flashing sheet centered over the seam. On the remaining three roofs only those seams with observed defects have been stripped as described above. The roofs that were selectively repaired have not exhibited a significantly higher number of seam related leaks than the remaining roofs in the study.

Documentation review also revealed that three of the EPDM membrane systems have had in excess of twenty-five percent of the partially cured neoprene flashing membrane replaced or reflashed. All three of these roofs were installed between 1978 and 1980 and the flashings most severely effected were exposed to south or west sun and when cured in place exhibited signs of stress and excessive thinning due to hand forming at horizontal to vertical transitions and by roller pressure at the transitions.

Review of project records revealed that 87 percent of the roof systems had had at least one claim filed for leak repairs and that 68 percent had multiple claims filed. Although there had been some roofs with several claims against warranty none have been replaced under the warranty.

Interviews of the party responsible for the maintenance of the roof systems were performed to gather unrecorded data relative to the roof systems performance and to ascertain the level of maintenance effort. Roof Maintenance Programs had been developed for 23 of the 36 roof systems by the designer which represented all roofs installed after 1980. All of the representatives interviewed reported performing roof maintenance on a regular basis, this was not reflected in the documented history of all the roofs and was not born out by the field evaluations. It appeared that the level of maintenance performed was highly variable but was relatively consistent for each multiple building owner.

FIELD EVALUATION

After completion of the document review and interview phases a field evaluation of the existing condition of each roof system was performed. For this evaluation a standard form was developed which could be utilized in checklist fashion, a copy of the Evaluation form is appended to this paper. In addition to evaluations the specific conditions of the roof membrane and membrane flashing, this field condition evaluation included assessment of sheet metal flashings, roof accessories and penetrations and rising walls above roof level.

None of the systems evaluated were found to be defect free. The extent of defects ranged from isolated punctures to extensive deterioration of flashing membranes and delamination of field fabricated membrane seams.

It was also apparent from the field evaluation that regular and periodic maintenance inspection and repair are critical to the performance of the single ply roof systems. Table 1 summarizes the defects observed. The defects noted during the field evaluation performed for this study all would have been found, during a properly performed maintenance inspection and could be repaired on a regular basis.

SUMMARY OF DEFECTS

Of note in regard to these defects are the following:

Insulation fastener membrane punctures only occured on assemblies with the membrane fully adhered directly over the mechanically fastened insulation. Design adjustment by installation of a layer of insulation in a mopping of hot asphalt (ASTM D312 Type III) eliminated the occurence of fastener puncture. Although not all punctures were cut open and examined it was typically noted that the punctures were due to fastener backout or improper fastener installation which resulted in the fastener head being above the stress plate.

- Embrittlement of elastomeric flashing membranes is most prevelant in older systems utilizing neoprene based flashing membranes. Defects typically noted on these flashings were checking, cracking and splitting. The membranes were also noted to be thinned and stressed where worked into plane transitions.

- Asphalt contamination of field sheets on both EPDM and PVC membrane systems was typically determined to be due to workmanship and can be avoided by careful application techniques and thorough surficial inspection prior to membrane application.

- Flashing membrane contamination due to bituminous residue on vertical surfaces was minimized by application of 3/8" thick pressure treated (preservative) plywood to all bitumen contaminated vertical surfaces prior to application of adhered membrane flashings. Application of the plywood also provided a surface with relatively constant absorbancy for application of adhesives and surface regularity to provide a finished appearance.

- Disbondment of fully adhered field membrane was observed on both EPDM and PVC membrane systems. Review of these systems revealed that disbondment occured at the insulation to membrane interface, typically attributable to workmanship during application and disbondment was also observed to occur on foam insulations between the insulation facer and foam surface. This latter type of disbondment has been attributed to inadequate bond development between the facer and foam during the manufacturing process which did not reveal itself until subsequent to adhesive application.

- EPDM membrane seam defects ranged from minor fishmouths and small openings at three way seam intersections (both field/field seams and factory/field seams) to total delamination of a seam. Seam defects were attributed to workmanship, adhesive failure, moisture contamination, and lack of step taper at factory seams.

- Contamination of field and flashing on both EPDM and PVC membranes due to oils and greases was observed predominantly below or adjacent to mechanical equipment (fan units, HVAC units, and kitchen exhaust units).

System performance was also generally affected by the level of maintenance performed by either the building staff or outside contract personnel. The most basic of maintenance deficiencies was a lack of periodic policing of the roof surface to remove accumulated debris and maintain a free flowing drainage system. It was also observed that lack of maintenance of related construction systems, including sealant joints and wall systems rising above the roof surface had deleterious effects on the performance of the single ply membrane system.

CONCLUSIONS

The analysis of single ply roof membrane systems performed to date in this study has revealed that the performance level has not been ideal. As with all exposed elements of a building it is imperative that a maintenance program be established and performed on a regular periodic basis to achieve the desired performance level. All of the roofs involved in this study were designed to exceed the minimum standards of the manufacturers current at the time of construction. Quality Control during construction included both part-time or full-time roof construction monitoring services as well as periodic and final inspections by representatives of the manufacturers and quantitative laboratory testing on a selective basis. Despite the level of design and quality control exerted during the construction phase none of the systems were found defect free and all require repair and maintenance.

As was previously noted some of the defects noted in older roofs included in this study have been eliminated by systems changes instituted by the manufacturers and design changes for the overall systems by the consulting firm.

It is anticipated that this performance study will be continued for the next several years with an expansion of the data base to in excess of 100 roof systems over five years in service life by 1991. At this same point in time ten percent of these roofs will have been in service beyond the term of the guarantees issued by the membrane manufacturer and the long term performance can be evaluated.

ACKNOWLEDGEMENTS

The author acknowledges with appreciation the contributions to this paper by Michael Maloney, Brian Hennessey and Bonnie Toma.

SYSTEM TYPE

DEFECT TYPE	BALLASTED ELASTOMERIC (6 TOTAL)	FULLY ADHERED ELASTOMERIC (25 TOTAL)	FULLY ADHERED PLASTOMERIC (5 TOTAL)
FIELD MEMBRANE			
Open/Disbonded Seams	N.O.	17	0
Fishmouths	N.O.	18	2
Wrinkles	N.O.	6	3
Disbonded/ Unadhered	N/A	18	5
Asphalt Contamination	N/A	10	0
Punctures			
Fastener	N/A	3	0
Ext. Damage	0	8	0
Contamination from External Source	2	10	2
Ballast Migration	6	N/A	N/A
FLASHING MEMBRANES			
Open Seams	4	17	0
Disbondment from Substrate	4	18	5
Wrinkles	2	19	3
Embrittlement/ Cracking	4	15	N/O
Punctures	3	12	1
OTHER			
INSULATION			
Fastener Backout	N/A	6	3
Delamination of Facers	N/O	4	N.O.
Damaged by Traffic	3	4	2
WOODBLOCKING			
Bowed/Deformed	2	10	1
Loose	3	14	3

N.O. = None Observed
N/A = Not Applicable

TABLE 1. - QUANTITY OF DEFECTS NOTED BY SYSTEM TYPE

CONTRIBUTING CAUSES

	D	MR	C	M
DEFECT				
FIELD MEMBRANE				
Open/Disbonded Seams		X	X	X
Fishmouths			X	
Wrinkles			X	
Disbonded/ Unadhered Membrane		X	X	
Asphalt Contamination			X	
Punctures				
Fastener	X	X	X	
Ext. Damage				X
Contamination from External Source				X
Ballast Migration	X	X		X
FLASHING MEMBRANES				
Open Seams		X	X	X
Disbondment from Substrate		X	X	
Wrinkles			X	
Embrittlement/ Cracking		X	X	
Punctures			X	X
OTHER				
INSULATION				
Fastener Backout		X	X	
Delamination of Facers		X		
Damaged by Traffic	X		X	X
WOODBLOCKING				
Bowed/Deformed		X	X	
Loose			X	

D = DESIGN   MR. = MANUFACTURER
C = CONSTRUCTION   M = MAINTENANCE
X = Contributing Factor

TABLE 2. - ASSESSMENT OF CAUSE TO DEFECT

	BALLASTED EPDM	FULLY ADHERED EPDM	FULLY ADHERED PVC
INSULATION			
Single Layer Urethane	4	8	-
Single Layer Urethane/Perlite Composite	-	14	3
Single Layer Isocyanurate	-	3	-
Two Layers Fiberboard Over Urethane	2	2	2
Two Layers Fiberboard Over Urethane/Perlite Composite	-	2	-
Two Layers Fiberboard Over Isocyanurate	-	1	-

NOTE: The Fiberboard layer of insulation on two layer
        installations was always installed using steep,
        asphalt under fully adhered roofs and loose laid
        on ballasted roofs.

TABLE 3. - INSULATION USEAGE BY ROOF SYSTEM TYPE

```
                           ROOFING SURVEY
DATE:                                              Inspected by:
Project #:
Project Name:
Location:
Owner:
Building Usage:
Roofing project completed:
Plan Attached:
Gen. Photo Attached:

1.0  DESIGN/INSTALLATION INFORMATION
1.1  Membrane Type:              Manufacturer:
     Sheet Thickness:            Dimensions:
Factory Seam Spacing:            Field Seam Spacing:
System   Ballasted:
                  Ballast Type:
                  Ballast Size:
                  Ballast Weight:


         Adhered:
             Adhesive type:
             Full/Partial:


     Seams - Adhesive (Type):         Weld(Solvent/Heat):
     Seam Sealant(Y/N):       Type:

1.2  Thermal Insulation
     R-value specified:       R-value used:      Tapered:     Uniform:
     Number of Layers:
     Upper Layer Type:
     Manufacturer:                           Thickness:
     Lower Layer Type:
     Manufacturer:                           Thickness:
     Attachment of upper layer:
     Attachment of lower layer:
     Mechanical Type:                  Length:             inches
     Manufacturer:                     Spacing:
     Adhesive Type:                    Coverage:
     Manufacturer:


1.3  Vapor Retarder
     Attachment - Loose:     Mechanical:     Adhesive:     None:
     Manufacturer:                           Type:

1.4  Deck Type - Metal:       Concrete:      Wood:
               Other:       Type:
     Manufacturer:                           Thickness:     inches
     Slope:

1.5  Flashings   Type:
                 Material:
                 Thickness:            inch

     Cants  Type:                      Size:
            Treated:                   Securement:

     Blocking Type:                 Size:
              Treated:              Fastener type:
              Spacing:              Size:
```

Single Ply Roof Membrane
Performance Study
Data Base Format
DB-1

SINGLE PLY SURVEY DATA SHEET

Project Name: DATE:

2.1.1 Walls	Y/N	Comments
1. Masonry		
A. Cracks		
B. Spalling		
C. Poor Mortar Joints		
D. Other		
2. Concrete		
A. CIP		
B. Pre-Cast		
C. Spalling		
D. Cracks		
E. Open Joints		
F. Other		
3. Siding		
A. Metal		
B. Wood		
C. Other		
4. Other		
A. Defects		
B. Other		

2.1.2 Roof Decks	Y/N	Comments
1. Visible from int.		
2. Rusting		
3. Spalling		
4. Cracking		
5. Buckling		
6. New Equipment or Penetration		
7. Other		

2.2.1 Roof Condition	Y/N	Comments
1. Discoloration		
2. Cracking		
3. Ponding water		
4. Debris		

Single Ply Roof Membrane
Performance Study
Data Base Format
DB-2

```
 5. Physical damage

 6. Punctures

 7. Plant growth

 8. Fungus

 9. Soft insulation

10.Other

2.2.2    Seams          Y/N              Comments
1. Open Joints

 2. Fishmouths

 3. Ridges

 4. Wrinkles

 5. Dried up Sealant

 6. Missing Sealant

 7. New Sealant

 8. Patches

 9. Other

2.2.3 Fully Ad. Mem.    Y/N              Comments
1. % Unadhered

 2. Insul. Fastener
    backing out

 3. Ridging

 4. Asphalt Contam.

 5. Bubbles

 6. Unadhered Insul.

 7. Other

2.2.4 Mech. Fast. Mem.  Y/N              Comments
1. Loose fasteners

 2. Fastener backout

 3. Other

2.2.5 Ballasted Membrane
1. % Ballast Disp.

 2. % Fracture

2.3      Condition of Flashing
1. Comment
```

Single Ply Roof Membrane
Performance Study
Data Base Format

DB-3

```
2.3.1  Base Flashing
1. Deterioration

2. Punctures

3. Attachment

4. Ridging

5. Sagging

6. Wrinkling

7. UV degradation

8. Unadhered

9. Open Joints

10.Fishmouths

11.Tears/splits @
   Nailer joints

12.Asphalt contamin.

13.Bubbles

14.Broken bubbles

15.Holes @ Fasteners

16.Loose Nailers

17.Splits @ metal
   flashings

18.Other

2.3.2  Counter Flashing
1. Punctures

2. Attachment

3. Rusting

4. Stains

5. Other

2.3.3   Coping
1. Open Fractures

2. Punctures

3. Attachment

4. Drainage

5. Other

2.3.4    Edge Metal
1. Gravel Stop
```

Single Ply Roof Membrane
Performance Study
Data Base Format

DB-4

2. Gutters

3. Other

4. Damage

5. Other

```
2.4.1   Exp Jt Covers
2.4.2   Walkways
2.4.3   Penetrations
2.4.4   Drains
2.4.5   Pitchpans
2.4.6   Caulking
2.4.7   Accessibility
2.4.8   Other (describe)
2.4.9   Previous Repairs
2.5     Comments
```

Single Ply Roof Membrane
Performance Study
Data Base Format

DB-5

Analytical Investigations of Roofing Systems

Heshmat O. Laaly[*]

PHOTOVOLTAIC SINGLE-PLY ROOFING MEMBRANES

REFERENCE: Laaly, H. O., "Photovoltaic Single-Ply Roofing Membranes," Roofing Research and Standards Development: 2nd Volume, ASTM STP 1088, Thomas J. Wallace and Walter J. Rossiter, Eds., American Society for Testing and Materials, Philadephia, 1990.

ABSTRACT: The past twenty years is considered to be a landmark and revolutionary advancement period for the progress and recognition of various classes of Single-Ply Roofing Membranes in Canada and the United States. During the same two decades, parallel and independently tremendous efforts and billions of dollars were allocated to generate electricity from solar radiation by advancing the level of understanding and implementation of photovoltaic science and technology. Prohibitively expensive, in 1969, the dream of landing on the moon and returning to Earth became a reality by combined efforts of the U.S. scientific community. The use of photovoltaic semiconductors on glass panels was a key factor for the success of this monumental project.

Seventeen years later in 1986, the author succeeded in bringing these two technologies together. The photovoltaic single-ply roofing membrane is a product which combines the usefulness and universal utilization of both science and technologies. This patented procedure, the basic research and development of its components, and the self sufficiency of the electrical energy generated is discussed. The testing and application standards proposed for this class of revolutionary, flexible and bi-functional roofing membranes, examples of their potentials for mass production and applications are also provided.

KEY WORDS: Photovoltaic, Electricity, Bi-Functional, Single-Ply, Roofing Membrane, Research, Application

[*] President, Roofing Materials Science & Technology, Los Angeles, CA 90035

INTRODUCTION

The main purpose of this paper is to introduce, demonstrate and provide an overview of a novel process to both the single-ply roofing industry as well as the photovoltaic research and development community. Every single-famly home could have a roofing material which is bi-functional, namely; structural waterproofing and provide free electricity utilizing the single-ply concept and photovoltaic scientific principles.

Because the technology is proven successful using protypes on a smaller scale, functional and tested in the laboratory, the negotiations for establishing a manufacturing plant are underway. An evaluation of full-scale model functionality would be the follow-up of this presentation.

The author has purposely discussed the photovoltaic aspects more in detail to provide a basis for roofing material research and standard development as related to photovoltaic single-ply roofing membranes.

THE SINGLE-PLY ROOFING MEMBRANE

Historical Background

Modified bitumens, rubbers and plastics are being extensively used in the manufacturing of membranes for roofing and waterproofing and are commonly termed as "Single-Ply Roofing Membranes." The major classification of the single-ply roofing membranes was conducted by the National Research Council of Canada (NRCC) [1] and the Canadian General Standards Board (CGSB) [2] since the early 70s. In the United States, the American Society for Testing and Materials (ASTM), Committee D08 [3], the Single-Ply Roofing Institute (SPRI) [4] and also the National Roofing Contractors Association (NRCA) [5] followed suit. Nowadays, the following categories are generally accepted, as indicated by product definitions such as those found in the Roofing Materials Guide [6]: prefabricated reinforced modified bitumens, vulcanized elastomers, non-vulcanized elastomers, and thermoplastic materials which are manufactured in thicknesses between 1 and 4 mm (0.040" and 0.120"), but so far, only their waterproofing properties have been utilized.

SCIENTIFIC PRINCIPLES OF PHOTOVOLTAICS

Historical Background

It all goes back to 1839, when the French scientist, Becquerel, noticed that when light fell on one side of a very simple kind of battery cell, it produced an electric current in a wire. For many years this phenomena was treated as trivial, for at this time there was no application. Twenty five years ago, as a matter of curiosity, scientists tried to shed light using scientific principles to improve this induced electrical current. Exactly 20 years ago, (June 20, 1969), the first Apollo spacecraft landed on the Moon. Refer to table below.

1839   Becquerel's discovery of the photovoltaic phenomena

1950	Bell Laboratory discovery of solar cell technology
1960	Market dominated by crystalline silicone (Space program)
1970	New thin film materials discovered, terrestrial PV R&D begun
1980	Technology mix of crystalline and amorphous silicone
	Remote applications become available
1990	Multi-junction thin film technology penetrated market
1986	Photovoltaic Single-Ply Roofing Membranes developed
1989	Photovoltaic Single-Ply Roofing Membranes patented [7]
2000	Forecasted high performance low cost PV technology
	Significant photovoltaic contribution by nation's utilities

By 1980, we had already used the solar heating and cooling principles extensively as indicated by T. Lucas [8], and by 1984, K. Zweibel and P. Hersch [9], laid down the basic photovoltaic principles and methods. The U.S. Department of Energy [10] has formulated a five-year research plan (1990-1995) dealing with the management and allocation of $55 million in funds for advancement and utilization of photovoltaic science and technology to help the US PV industry transfer the research results from the laboratory into production processes that yield stable, high-efficiency, low-cost cells that can reach the U.S. Department of Energy goals of 12¢/kWh in the mid 1990s.

In collaboration with Sandia Research Laboratories [11], the development and compilation of a handbook for stand-alone photovoltaic systems has been devised. In addition, the National Electrical Code Handbook 1990[12], has devoted an entire section on photovoltaics dealing with all facets of application used in conjunction with electrical household use as indicated on pages 962 to 974.

Today's understanding of PV is as follows: Light is converted into electricity by the photo-sensitivity of certain semiconductor materials. When photons of light strike such materials, they induce a random movement of electrons. A difference in electrical potential between the top and bottom layers of the semiconductor causes an electric current to flow.

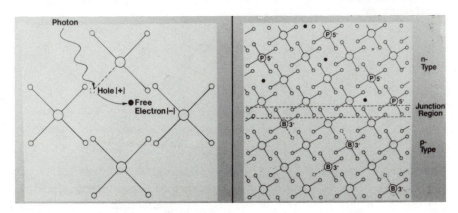

FIGURE 1 -- (A) Photon separates electron (negative charge) from its electromagnetic hold (positive charge). (B) Electrical potential results from "doping" silicon with impurities. P^{5+} (Phosphorus) makes n-type silicon, while B^{3+} (Boron) makes p-type silicon.

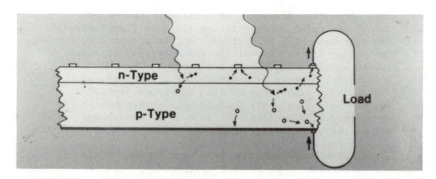

FIGURE 2 -- Radiant energy penetrates the solar cell, causing excess electrons to accumulate in n-type area and excess holes to accumulate in p-type area. The circuit is completed when a load is applied and charges are drained off.

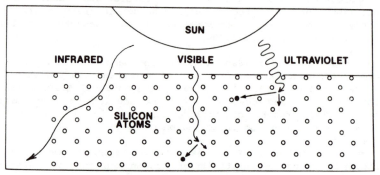

FIGURE 3 -- Low energy infrared light cannot separate an electron from its hole, and therefore will not generate electricity. High energy ultraviolet light strikes electrons too violently, which generates unwanted heat. Only the middle range is converted to electricity effectively.

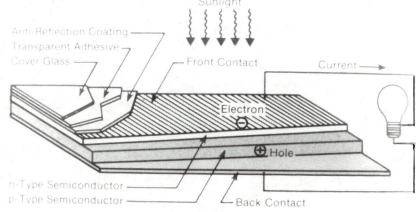

FIGURE 4 -- In a typical cell, sunlight frees electrons and internal voltage is created that drives current through an external circuit.

Different types of semiconductors have different energy gaps (band gaps) corresponding to the distribution of photons in the solar spectrum. The multi-layer semiconductors are called cascade cells and can be utilized to optimize the electrical energy conversion. Tiny amounts of specified impurities, called dopants, are added to the top and bottom layers of the cells enhancing the positive and negative charges without changing the positive or negative value of the semiconductor material.

Amorphous and crystalline silicon cells have gained world-wide interest for economic reasons, while other cells made of Gallium Arsenide (GaAs), Copper Indium Diselenide ($CuInSe_2$) (CIS) are more expensive and have found greater use in space exploration.

THE SUN AND ROOFS

From Foe to Friend

It is well established that solar radiation has extreme deteriorating effects on organic materials and particularly, roofing materials. In the wide spectrum of sunlight, wave lengths of the ultra violet (UV) in terms of kilo-calories are in the vicinity of molecular bond energy. For example, the carbon-bonded atoms, (C-C), (C-H), (C-O), (C=O), (C=S), require different quantums of energy for their formation and decomposition. The conventional roofing material, bitumen, which once was the major product for roofing, upon exposure to ultra violet radiation hardens rapidly, and due to loss of volatile components, cracks (alligatoring) rapidly. For this reason, protection from solar radiation through utilization of layers of gravel, granules, or other protective materials are essential.

As far as the single-ply roofing membranes (modified bitumen, thermoplastic, elastomeric) are concerned, during manufacturing, some sort of UV absorbers are incorporated in the formulation, to divert decomposition from the resins or polymers. This is not only a cost factor, rather it has been carefully selected and evaluated in order that a long service life and satisfactory performance be attained.

The sun is the governing body of our existence in every aspect. It would be of tremendous importance if we succeed in converting this energy source to the benefit of all mankind. Reduction in pollution, less consumption of non-renewable energy and convenience are just a few examples of the advantages and benefits.

Most of the statistics for the sun listed below represent the best approximations based on available scientific data and calculations: Solar power delivered to Earth daily: 85 trillion Kilowatts. Note: The sun delivers enough energy in 15 minutes each day to supply the world's total energy needs for a year. Solar energy per square foot in the United States: 75 to 104 BTUs per hour in summer, 21 to 46 BTUs per hour in winter. Distance of sun from Earth: 92,913,000 miles. Difference in distance of sun from Earth between January and June: 3,069,000 miles. The sun is almost 110 times greater in diameter than Earth. Volume of the sun compared to Earth: 1,300,000 : 1. Orbital velocity of Earth around sun: 18.5 miles per second. Surface temperature of sun: 10,800°F (or 6,000°C). Interior temperature of sun: 27,000,000°F (or 15,000,000.C). Wavelength of sunlight: 0.3 to 3.0 nanometers (microns). Note: 1 namometer = 1 millionth of a millimeter. Area of roof coverage in the United States (alone), (population of 250 million): 21,500,000,00 square feet (or approximately 860 square feet per person). The energy in sunlight: approximately 1,358 Watts per square meter. (930 Watts at sea level).

According to accurate measurements and findings of the U.S. Department of Energy, depending on the geographical area, weather conditions and seasonal variations, the Earth is radiated with an average of between 2.5 and 6 kWh per square meter per day and the corresponding figure for the United States is an average of between 2.5 and 7.5 kWh per square meter per day.

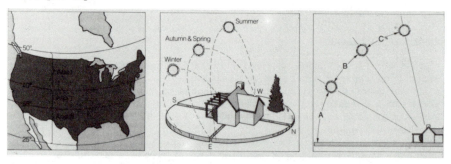

FIGURE 5 -- Shade is cast at various angles, depending on time of year and location. Find your location on map above and refer to TABLE 1 for sun angles on your property. [13]

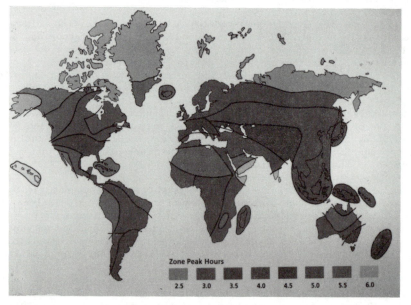

FIGURE 6 -- World-wide peek hour zones with insolation expressed in kWh/m^2.

TABLE 1 -- Seasonal Sun Angles [13]

Season	Sun's Position/Hours of Daylight (see map above)		
	AREA I	AREA II	AREA III
A Noon, DEC 21	21°/8 hrs.	29°/ 9 hrs.	37°/10 hrs.
B Noon, MAR 21 & SEP 21	45°/12 hrs.	53°/12 hrs.	60°/12 hrs.
C Noon, JUN 21	69°/16 hrs.	76°/15 hrs.	83°/14 hrs.

SERI [14] has developed a computerized method of calculating Annual Mean Daily Direct Beam Solar Radiation expressed in $kWh/m^2/day$ as a way of graphically expressing insolation values for geographical areas of the U.S.

The purpose of this discussion is to emphasize the importance to the scientific community worldwide that this tremendously valuable resource should not be neglected any longer and the surface of the roof would be a most logical area on which the photovoltaic energy can be harnessed for individual consumption regardless of where the building or shelter is located.

BASIC RESEARCH AND DEVELOPMENT OF PHOTOVOLTAICS

PV Cells, Modules and Arrays

Kenneth Zweibel [15] indicates that photovoltaic cells are semiconductor devices that transform light energy continuously into direct-current electricity. The smallest unit photovoltaic device is called a cell. Several electrically interconnected cells in framed glass panels are called a module (their size is usually conveniently 60 cm. x 120 cm. and may range in area from about 100 to 10,000 sq. cm). When photovoltaic devices are used to generate power, many modules are wired together into a photovoltaic system called an array. Multiple arrays are combined together to form what is referred as a photovoltaic farm. This type of farm is usually installed by a power company to supplement the conventional electrical generation during high-energy load periods.

The principles by which solar cells convert light into electricity are easy to understand. Semiconductors can have their electrical properties dominated by free charges of either the negative (electrons) or positive (holes in the valence band) type. When two semiconductors dominated by opposite electrical charges are in contact with one another, free charge leaks across their common boundary and becomes fixed as ions in the regions adjacent to the common interface. The fixed but opposite ions produce at the interface a local region with an electric field. That electric field sends free electrons one way and free holes the other. Normally, no current flows in this diode-like device.

The success of photovoltaic devices depends partly on achieving good sunlight-to-electricity conversion efficiencies. The need for less expensive cells has spawned the so-called thin-film and concentrator technologies. These are strategies for producing cells inexpensively, either by making the cells so thin that they require little cost for raw materials or processing, or by using cheaper lenses to concentrate sunlight on small-area cells. Those two approaches form the bulk of the efforts to design cost-effective photovoltaic systems for generating commercial electricity.

A final parameter of importance for photovoltaic cells is long-term durability. Most photovoltaic arrays for making electricity will require lifetimes of 20 to 30 years. Actually, existing photovoltaic demonstration projects have shown that megawatt arrays can work for years almost without human supervision. The potential for 30-year, maintenance-free electricity is a powerful plus for photovoltaics.

Solar cells are built in several designs. The design of photovoltaic cells varies to counteract the limitations in the cell material. The four basic designs are: homojunction cells (using crystalline silicon); heterojunction cells ($CuInSe_2$, CdTe, or GaAs); p-i-n cells (amorphous silicon); and cascade combination of these structures. In a cascade structure, high energy photons are absorbed in a cell that is tuned to high energies, and low-energy photons in a cell that is tuned to low energies. This results in more efficient use of the solar spectrum. Each component cell uses essentially all of the light above its band gap, (the range of the light spectrum converted to electricity) thereby producing power proportional to the number of photons times a voltage that is a fraction of its bandgap.

In September, 1985, the Construction Specifier Magazine [16] reported that ARCO Solar (a subsidiary of Atlantic Richfield), in partnership with the U.S. Department of Energy's (DOE) Solar Energy Research Institute (SERI) [14], reportedly achieved a 30 percent increase in the performance of a low-cost one-square-foot solar panel. ARCO's achievement, coupled with the cost efficiency of large-scale production of solar panels, reportedly has the potential of bringing solar electricity into the reach of U.S. consumers at or below 13 cents per kilowatt-hour. The new panel converts 11.2 percent of the sun's energy received to electricity and is made from an innovative semiconductor material called copper indium diselenide, or CIS ($CuInSe_2$). The DOE goal for solar panel performance is 15 percent conversion efficiency.

DEVELOPMENT OF PHOTOVOLTAIC SINGLE-PLY MEMBRANE

Rationale and Specific Design Considerations

As early as the 1950s until 1986, the most common substrate for photovoltaic cells was glass. The glass had to be treated specially for this purpose. All depositions had to be transparent. During the mid 80s, the Japanese PV industry succeeded in producing photovoltaic tiles, which had to be interconnected to obtain the electrical output of all individual tiles combined.

In the United States, the photovoltaic films were deposited on flexible stainless steel and polyamide film.

The author's innovative process has utilized the chemical principles and extensive knowledge and experience of single-ply roofing membrane science and technology and developed a technique which has eliminated all the above-mentioned limitations.

In contrast to conventional PV modules, no glass substrate is used. Any of the already available single-ply roofing membranes (thermoplastic, elastomeric or modified bitumen could successfully be used for their waterproofing properties and as a substrate for photovoltaic cells.

The details, parameters and selection of components are far beyond the scope and limited pages of this paper, however, the process to incoporate a rigid, inorganic semiconductor with a flexible, single-ply roofing membrane has been possible utilizing the following conditions:

- Using cells of a smaller dimension,
- Allowing space between cells with a slight indentation to permit rolling and unrolling of the membrane on a 5-to 6-inch diameter core,
- Selecting connecting wires and techniques with appropriate durability and flexibility to allow for bending several times without breakage,
- Selecting suitable cell thicknesses which reduces breakage factor when sustaining foot traffic during application or maintenance in addition to use of a cushioning layer (pottant).

It has to be mentioned that in normal roofing application practice the membrane may undergo rolling or unrolling two or three times from the factory to actual placement on the roof and it is unlikely that the membrane will require an inherent amount of repeated bending.

MODULE SUNSIDE	LAYER DESIGNATION	FUNCTION
	SURFACE 1) MATERIAL 2) MODIFICATION	• LOW SOILING • EASY CLEANABILITY • ABRASION RESISTANT • ANTIREFLECTIVE
	TOP COVER	• UV SCREENING • STRUCTURAL SUPERSTRATE
	POTTANT	• SOLAR CELL ENCAPSULATION
	SPACER	• ELECTRICAL ISOLATION • MECHANICAL SEPARATION
	SUBSTRATE	• STRUCTURAL SUPPORT
	BACK COVER	• BACKSIDE MECHANICAL PROTECTION • BACKSIDE WEATHERING BARRIER
(A)		

FIGURE 7(A) -- Please see explanation on next page

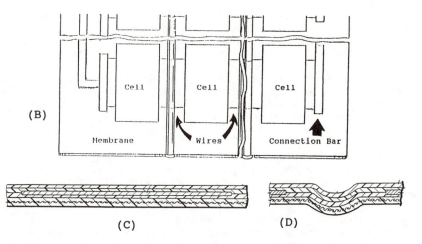

FIGURE 7 -- The above figure (A) shows an example of the most common components and different layers of the photovoltaic cell. The schematic drawing (B) shows the electrical wiring of the fabricated PV Single Ply Membrane prototype. The cross-section (C) shows a schematic representation of the membrane layers and the grooved area (D) shows provision of additional flexibility for rolling and unrolling of the membrane during application. The membrane fabrication is based on hot vacuum oven technology.

This invention consists of bonding photovoltaic solar cells through a heat vacuum lamination process to a flexible roofing membrane. The reduced pressure and varied temperature range are a function of processing speed and the type of single-ply membrane material incorporated. The solar cells are electrically interconnected into arrays in order to generate electrical power which possesses the desired voltage and current characteristics. After lamination, these solar cells are encapsulated in a transparent padding material, and further protected by a transparent plastic overlay, specially selected for its durability and resistance to dirt.

The roofing membrane provides a base to which the solar cells are laminated and environmentally protects the structure on which it is applied.

The primary advantages of this new class of revolutionary and bi-functional roofing membranes are as follows:

- provides environmental protection to the structure where it is installed;

- collects solar energy and converts it into electrical power with the desired characteristics;

- generates usable electrical power when installed on a typical residential or commercial structure, in a magnitude of thousands and millions of watts respectively, since the amount of electricity generated is directly proportional to the roof surface area;

- can be installed on roof substrates of any type configuration, i.e., gradual slope, sharp pitch, flat, or in any combination thereof using conventional fully-adhered or mechanically fastened techniques;

- remains flexible after manufacture, including factory encapsulation (lamination) of the solar cells, so that it is capable of being rolled up for transport and unrolled at the construction site;

- critical electrical interconnections to the solar cells are performed at the factory under controlled conditions, as opposed to being performed at the construction site;

- can be manufactured in roll-form to any desireable or convenient width and length;

- enables installation at the construction site by conventional single-ply applicators with a few hours of additional training;

- requires little or no maintenance after the system is installed;

- excessive generated (unused) electricity can be stored in suitable battery packs with a service life up to 10 years;

- generated DC current can be converted to AC and all electrical engineering standards and codes of practice are applicable;

- no changes to existing electrical wiring is required. Connections are made at the power panel or fuse box to interface interchangeably between the two power sources; and,

- can be used as a hybrid system along with existing power utility services or serve as sole source of alternate energy.

TESTING AND EVALUATION

Durability, Functionality & Performance Criteria Considerations

Every PV cell's efficiency is judged by a ratio of insolation and electricity in direct current (dc) generated per given surface area and given time.

Amorphous silicon cells interconnected with wires and contact bars were used. Other types of PV cells can be used equally well. Refer to the scales in FIGURE 8 for actual dimension of the cells.

As a pilot project, three pieces of photovoltaic single-ply roofing membrane prototype were fabricated under contract at ISET Laboratories [17] in Los Angeles PV R&D facilities; Two pieces (30 x 60 cm) each and one piece (40 x 80 cm). The results of two of the prototypes' electrical output are reported as shown in FIGURE 9.

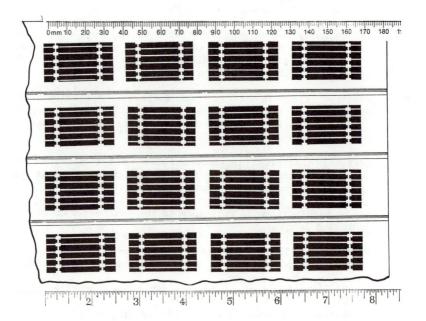

FIGURE 8 -- The actual cell configuration and wiring of the two small reinforced PVC-based membranes with International and U.S. scales.

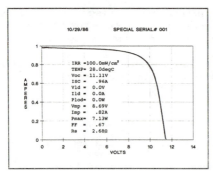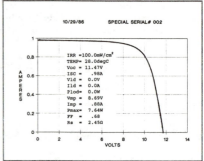

FIGURE 9 - The I-V Curves obtained.[17]

As shown in FIGURE 9, the I-V curves represent the electrical DC current outputs obtainable from two of the three prototypes developed each measuring two square feet in surface area.

Table 2 shows the actual values of testing result output as analyzed by ISET Laboratory equipment on October 29, 1986.[17] Comparison of the results from the data obtained show uniformity of output in the form of maximum voltage versus amperage from both prototypes.

TABLE 2 -- Voltage Output Analysis

	Serial #1	Serial #2
Irradiation Rate (IRR in mW/cm^2)	100.0	100.0
Temperature (in °C)	28.0	28.0
Voltage Open Current (Voc in Volts)	11.11	11.47
Power Short Circuit (Isc in Amps)	.96	.98
Load Voltage (Vld in Volts)	0.0	0.0
Load Current (Ild in Amps)	0.0	0.0
Load Power (Plod in Watts)	0.0	0.0
Voltage Maximum Power (Vmp in Volts)	8.69	8.69
Current Maximum Power (Imp in Amps)	.82	.88
Maximum Power (Pmax in Watts)	7.13	7.64
Fill Factor (FF)	.67	.68
Circuit Resistance (Rs in Ohms - Ω)	2.68	2.45

The normal durability for single-ply roofing membranes are generally between 10 and 20 years when it is exposed to adverse environmental conditions. With the inclusion of the photovoltaic components and transparent top layer, the actual membrane is protected from pollutants, moisture, and to a large degree, UV radiation. In other words, even after 20 years of providing free electricity, the waterproofing membrane is practically unweathered and will continue to function.

APPLICATION OF THE PHOTOVOLTAIC MEMBRANE

Differences from Conventional Membranes and Bi-Functionality

At this early stage of the art of photovoltaic single-ply roofing membrane technology, the author believes that the application techniques will be basically similar to those used for conventional single-ply membranes. However, an ASTM Committee [3] should be formed in order to decide which one of the abbreviations, such as PVM, PVRM, PVSRM or PVSPRM are to be used for this new type of membrane, develop appropriate material standards, and to highlight and specify the needed special considerations for application of each class of these PV Membranes. The topics to be covered are practically all those discussed in existing application standards with emphasis on safe handing; soft-soled shoes, use of walkways, lap joint techniques, substrate issues, wiring, snow removal in winter, yearly cleaning with detergent, heat dissipation, electrical storage devices and systems, power inverters and voltage regulators just to name a few. Fortunately a wealth of data has already been generated by IEEE/PVSC [18], and the above-mentioned organizations which has to be considered. The golden rule of 1:24 to 1:48 minimum slope ratio or 41.5 to 20.75 mm/m (2°23' to 1°12' approximate degree slope) on a flat roof is highly recommended if not mandatory. The minimum advantage of assuring a non-ponding roof would be the avoidance of dirt accumulation, higher sunlight penetration and consequently the optimization of electrical output of the membrane.

The details of application of conventional single ply have to be slightly modified, whereby, care has to be taken that cells and connections are not damaged by sharp objects or allowing unnecessary traffic on the membrane during application. Where traffic becomes necessary, a temporary plywood walkway must be used, the provision of a narrow and permanent walkway for maintenance or service personnel is recommended and preferably, air conditioning installations, ductwork and other details should be completed prior to application of the membrane.

For estimation of PV system requirements, calculate the PV array area in square meters that you would need to satisfy your average daily electric requirement:

$$\text{Membrane Area } (m^2) = \frac{\text{Average daily load requirement in (kWh/day)}}{\text{Insolation (kWh/m}^2\text{/day)} \times \text{PV System efficiency}}$$

FUTURE WORK

Since the functionality of photovoltaic single-ply roofing membranes has been proven and the production of photovoltaic semiconductors has reached economic feasibility levels, the assistance of governmental, industrial and financial institutions are necessary. In the meantime, the author is in the process of constructing full-scale models on three single-family homes in different geographical locations with varying insolation values. This program is in conjunction with large-scale manufacturing research and development. Future years will be devoted towards in-depth studies to increase the level of efficiency and durability of the PV Membrane and reduction of costs.

SUMMARY/CONCLUSION AND FUTURE TRENDS

How we proceed to develop photovoltaic energy inevitably and intimately relates with the extent of universal application. The future depends on our ability to reach imaginative solutions as the generation of power is decentralized.

The audience of roofing scientists gathered in this Symposium and researchers acting in an official capacity such as ASTM [3] are the most appropriate levels through which the issue of PV Single-Ply research and standards should be implemented.

The stakes are very high, and we cannot wait until every aspect of the technology is completely ready for general consumption. Photovoltaic power and the changes that it will bring are inevitable. We would be well advised to deal with these realities affirmatively and deal with them now.

[19] We are face to face with a watershed decision on a new energy technology. Photovoltaic technology is no longer exotic, casting a distant rosy glow over the twenty-first century. We must begin to treat it as a practical here-and-now reality to be integrated fully into our

decisions about energy supplies.

The author has provided the necessary data to indicate the potentials of global utilization and gained environmental improvements and envisages that in the near future, stand-alone photovoltaic systems will become an essential part of various shapes and types of roofs. In other words, the conventional single-ply roofing membranes will be advanced to a higher level of universal utilization by becoming bi-functional; namely waterproofing of the building and silent provider of free electricity sufficient for all the needs of every household.

A GLOSSARY OF PHOTOVOLTAIC TERMS

AC Current
Electrical current in which the flow is reversed at frequent intervals, 120 times per second or 60 cycles per second, as used in commercial grid power in the United States and Canada. Opposite of Direct Current (DC)

Amorphous
The condition of a solid in which the atoms are not arranged in an orderly pattern; not crystalline.

Balance of System (BOS)
The parts of a photovoltaic system other than the array. Balance of System components include switches, controls, meters, power-conditioning equipment, supporting structure of the array and storage components, if any. The cost of land is sometimes included when comparing total system costs with the cost of other energy sources.

Band Gap Energy
The amount of energy (in electron volts) required to free an outer shell electron from its orbit about the nucleus to a free state and thus to promote it from the valence level to the conduction level.

Boron (B)
A chemical element, atomic number 5, semi-metallic in nature, used as a dopant to make p-silicon.

Conversion Efficiency (cell)
The ratio of the electric energy produced by a solar cell (under full sun conditions) to the energy from sunlight incident upon the cell.

Diffuse Insolation
Sunlight received indirectly as a result of scattering due to clouds, fog, haze, dust or other substances in the atmosphere. Opposite of direct insolation.

Direct Current (DC)
Electric current in which electrons are flowing in one direction only. Opposite of Alternating Current (AC)

Direct Insolation
Sunlight falling directly upon a collector. Opposite of diffuse insolation.

Dopant
A chemical element added in small amounts to an otherwise pure crystal to modify its electrical properties. An n-dopant introduces more electrons that are required for the perfect structure of the crystal. A p-dopant creates electron vacancies in the crystal structure.

Energy Payback Time
The time required for any energy-producing system or device to produce as much useful energy as was consumed in its manufacture and

construction.

Fill Factor (FF)
The ratio of the maximum power a photovoltaic cell can produce to the theoretical limit if both voltage and current were simultaneously at their maximums. A key characteristic in evaluating cell performance.

Insolation
Sunlight, direct or diffuse. Not to be confused with **insulation**. See diffuse insolation, direct insolation

Inverter
Device that converts DC to AC. See Rectifier.

I-V Curve
A graphical presentation of the current versus the voltage from a photovoltaic cell as the load is increased from the short circuit (no load) condition to the open circuit (maximum voltage) condition. The shape of the curve characterizes cell performance.

Load
The amount of electric power being consumed at any given moment. Also, in an electrical circuit, any device or appliance that is using power. See base load; peak load

n-Silicon
Silicon containing a minute quantity of impurity, or dopant, such as phosphorus, which causes the crystalline structure to contain more electrons than required to exactly complete the crystal structure. There is no electrical imbalance, however. Opposite of p-silicon.

Peak Load
The maximum load, or usage, of electrical power occurring in a given period of time, typically a day.

Phosphorus (P)
A chemical element, atomic number 15, used as a dopant in making n-silicon.

Photon
A particle of light, which acts as an indivisible unit of energy; a quantum or corpuscle of radiant energy moving with the speed of light.

Photovoltaic (PV)
Pertaining to the direct conversion of light into electricity.

Photovoltaic Array
A collection of PV modules, electrically wired together and mechanically installed in their working environment.

Photovoltaic Cell
A device that converts light directly into electricity. All photovoltaic cells produce direct current (DC).

Photovoltaic Module
The smallest replaceable unit in a PV array. An integral encapsulated unit containing a number of PV cells.

Photovoltaic System
A complete set of components for converting sunlight into electricity by the photovoltaic process, including array and balance-of-system components.

p-Silicon
Silicon containing a minute quantity of impurity, or dopant, such as boron, which provides insufficient electrons to exactly complete the crystal structure. There is no electrical imbalance, however. Opposite of n-silicon.

Rectifier
A device that converts AC to DC. Compare inverter.

Semiconductor
Any material that has limited capacity for conducting an electric current. Certain semiconductors, such as silicon, gallium arsenide, Copper Indium Diselenide (CIS) and cadmium sulfide, are uniquely suited to the photovoltaic conversion process.

Silicon (S)
A chemical element, atomic number 14, semi-metallic in nature, dark gray, an excellent semiconductor material. A common constituent of sand and quartz (as the oxide). Silicon crystallizes in face-centered cubic lattice-like diamond.

Volt, (V)
A measure of the force or "Push" given the electrons in an electric circuit; a measure of electric potential. One volt produces one amp of current when acting against a resistance of one ohm.

Watt
A measure of electric power or amount of work done in a unit of time. One amp of current flowing at potential of one volt produces one watt of power.

REFERENCES

[1]   National Research Council of Canada (NRCC), Montreal Road, Ottawa, Ontario, CANADA K1A0R6, (613) 993-9101 (General Inquiries), Institute for Research in Construction (IRC) - Building Materials Section, (613) 993-1596

[2]   Canadian General Standards Board (CGSB), Ottawa, Ontario, CANADA K1A0S9, (800) 267-8220 (General Inquiries)

[3]   American Society of Testing and Materials (ASTM, Committee D-8), 1916 Race Street, Philadelphia, PA 19103, (215) 299-5400

[4]   Single-Ply Roofing Institute (SPRI), 1800 Pickwick Avenue, Glenview, IL 60025, (312) 724-7700

[5]   National Roofing Contractors Association (NRCA), One O'Hare Centre, 6250 River Road, Rosemont, IL 60018, (312) 318-NRCA

[6]   "Commercial, Industrial & Institutional Roofing Materials Guide", Vol. 9, August 1986, NRCA Publication [5].

[7]   Laaly, H.O., et al, "Photovoltaic Cells in Combination with Single-Ply Roofing Membranes", U.S. Patent Number 4,860,509, August 29, 1989

[8]   Lucas, T., "How to Build a Solar Heater", A Herbert Michaelman Book, Crown Publishers, Inc., New York, 1980

[9]   Zweibel, K., Hersch, P., "Basic Photovoltaic Principles and Methods", Van Nostrand Reinhold Co., New York, 1984

[10]   U.S. Department of Energy, "Five Year Research Plan 1990-1995 Photovoltaics: USA's Energy Opportunity - National Photovoltaic Program", DOE/CH 10093-7, May 1987

[11]   Sandia National Laboratory, Photovoltaic Design Assistance

Center, "*Stand-Alone Photovoltaic Systems*", SAND 87-7023, Albuquerque, New Mexico, April 1988

[12]   National Fire Protection Association, "*The National Electrical Code®-1990*", 2 volumes, 1 Batterymarch Park, P.O. Box 9101, Quincy, MA 02269-9101, Fifth Edition, 1990

[13]   Sunset Books and Magazines, "*Patio Roofs and Gazebos - Design Ideas, Plans, Installation Techniques*", Lane Publishing Company, Menlo Park, CA 94025, Second Printing, May 1989

[14]   Solar Energy Research Institute (SERI), A Division of Midwest Research Institute, U.S. Photovoltaic Patents: 1951-1987, 1617 Cole Boulevard, Golden, CO 80401

[15]   Zweibel, K., "*Photovoltaic Cells*", Chemical & Engineering News, July 7, 1986, (pp. 34-48)

[16]   Construction Specifier Magazine, Construction Specifications Institute (CSI), 601 Madison Street, Alexandria VA 22314-1791, (703) 684-0300. For more information, contact Sy Morgansmith, Communications Manager, at (303) 231-7683.

[17]   International Solar Electric Technology, Inc. (ISET) 6635 Aviation Blvd., Inglewood, CA 90301, (213) 216-4427

[18]   Institute of Electrical & Electronic Engineers (IEEE), Photovoltaic Specialists Conference (PVSC), 345 E. 47th Street, New York, NY 10017, 1988 Chairman: Joseph Wise, (513) 255-6235

[19]   Maycock, P. D., "*A Guide to the Photovoltaic Revolution*", Rodale Press, Emmaus, PA, USA, 1985

James R. Wells and Jeffrey A. Tilton

A PRELIMINARY INVESTIGATION OF THE BIAXIAL LOAD-STRAIN
TESTING OF SELECTED ROOFING SHEET MATERIALS

REFERENCE: Wells, J. R. and Tilton, J. A., "A
Preliminary Investigation of the Biaxial Load Strain
Testing of Selected Roofing Sheet Materials,"
Roofing Research and Standards Development: 2nd
Volume, ASTP 1088, Thomas J. Wallace and Walter J.
Rossiter, Eds., American Society for Testing and
Materials, Philadelphia, 1990.

ABSTRACT: The roofing industry has seen increasing
efforts to move toward performance standards rather
than prescriptive standards for roofing materials.
In order to better understand the performance of the
materials, testing should represent the material
service environment as closely as possible. When in
service, roofing materials are stressed biaxially.
Therefore, load-strain testing to evaluate
performance of roofing materials should be biaxial.
Selected membrane sheet materials are tested, and
the results reveal significant reductions in
elongations to failure as well as failure strength.

KEYWORDS: biaxial,load,strain,testing,roofing

INTRODUCTION

The current decade of change in the commercial roofing
industry has witnessed the introduction of multiple new
sheet materials for roofing systems. Many of these
materials have physical properties which are markedly
different from the more traditional roofing materials. In

Dr. Wells and Mr. Tilton are researchers at the Owens-
Corning Fiberglas Technical Center, 2790 Columbus Rd., Route
16, Granville, OH 43023; Dr. Wells is a Senior Engineer,
and Mr. Tilton is a Senior Technician.
117

particular, many single ply roofing materials have significantly lower tensile strengths than traditional built-up roofing membranes, together with significantly greater elongation. The performance of many of these alternate materials do not meet the standards which were developed for the traditional higher strength materials. Further, if the lower strength/higher elongation materials do not conform to the traditional standards, the question arises as to what alternate standards are most appropriate for these materials. This is the current state of the roofing industry; for traditional high strength roofing membranes, BSS 55 from the National Bureau of Standards remains the most accepted set of recommended performance criteria, and for alternate membrane materials, new independent standards are developing.

While it is certainly possible to have separate performance criteria for each type of roofing system, many researchers have recognized that having recommended criteria that are valid for all types of roofing systems would be valuable. Efforts could be focused on the essentials of what is required for roofing membranes to perform well, regardless of the type. Focusing on strain energy at failure was such an attempt. Although it has not proven to be an unambiguous common measure thus far, further work may well develop this concept into a common criteria for all roofing systems. But whether or not strain energy or cyclic changes in strain energy become criterion of choice, efforts to identify common criteria should continue. The questions remain. What are the important performance criteria? What are the essential similarities and differences between the single ply and traditional materials? To what extent do common principles govern their behavior? To answer these questions, the properties of the individual materials and their service environment must be understood and analyzed as rigorously as possible.

This study focuses on the tensile and elongation properties of roofing sheet materials in the conditions normally experienced in service, conditions of biaxial loading. Even though it is well recognized that in actual service roofing membranes are subject to biaxial loading, either isotropically or to varying degrees, essentially all testing of roofing materials has employed only uniaxial tests. The degree of appropriateness of this testing may well be material dependent, but the most valuable data and the most valid correlation will be possible using data that is obtained via testing which properly represents the service environment. A preliminary investigation of the biaxial load-strain behavior of several common roofing sheet materials is reported in this study; two APP modified bitumens, two EPDM sheets, and one PVC.

BIAXIAL TESTING

Although biaxial testing is clearly more representative of membrane material behavior than uniaxial testing, comparatively little biaxial testing has been done owing to the added complexity of the tests. However, though not widely used, they have been well characterized and used more commonly with tests of coated fabrics or metal plates. J. C. Radon, et al[1] describes biaxial testing of homogeneous materials and gives good illustrations of sample preparation. Works by R. Mott, G. Huber, and A. Leewood[2], R. B. Testa and W. Boctor[3], and H. W. Reinhart[4] address testing of coated fabrics and are more applicable to testing of roofing materials.

Biaxial test methods are of three principle types; bursting tests, cylinder tests, and plane biaxial tests. The bursting tests are suitable for only small specimens and may yield inaccurate results for anisotropic reinforced or woven fabrics. The cylinder tests and plane biaxial tests are more useful. Mott, et al[2], used the cylinder method with good success in characterizing permanent fabric roof materials. Reinhardt[4] discusses each of these tests and states that the best measurements for stress and strain can be obtained from plane biaxial tests. All testing reported in this study employed plane biaxial testing.

The sample loads and grip displacements were measured using load cells and LVDTs. The deflection of the biaxial area was either measured by hand with a linear scale on each axis, or additional LVDTs were mounted on the sample surface, as in Figure 19. For high elongation materials tested at room temperature, the most convenient method for measuring elongation was to measure the biaxial area at selected load increments. For low elongation materials, which require greater accuracy in the elongation measurements, and for all materials tested with the environmental chamber in use, additional LVDTs were attached to both sides of the biaxial area on each axis. For all the materials, a small spot of hot glue was sufficient to attach the LVDT mounts to the sample surface.

EQUIPMENT

The testing was done on a custom built biaxial test machine referred to as the Biaxial Roof System Tester (BRST), located in the research facilities of the Commercial Roofing Laboratory at the Owens-Corning Fiberglas Technical Center in Granville, Ohio. The BRST is a large cruciform test frame which will accommodate samples as large as 4 meters by 4 meters (13 x 13 feet) in length. The sample width on each axis, and consequently the size of the biaxial test area, is determined by the size of grips in use. For large scale roof system tests, widths up to 52 cm (20.5 in) have been used, yielding a 52 cm by 52 cm (20.5 x 20.5 in) test biaxial area.

In the current testing, a much smaller configuration of the tester and smaller test sample is used. The test samples are 10.2 cm (4 in) in width on each axis, resulting in a 10.2 cm (4 in) square biaxial area. The distance between the grips (gage length) is varied depending on the elongation of the test material. Since the tester was originally designed for use with materials of relatively low elongation, the total travel of the hydraulic cylinders is limited to a designed maximum distance of 23 cm (9 in). Consequently, the gage length must be chosen with regard to this extension limit and the anticipated elongation of the material to be tested.

The BRST control and data acquisition system was custom designed using standard component modules. A Hewlett Packard Series 80 Personal Computer is responsible for directing all aspects of tester operation; specification of displacement or load histories, temperature histories, and data acquisition. The dynamics of the system hydraulics are controlled by four MTS model 406 controllers interfaced to the computer. Analog to digital and inverse conversions and all signal conditioning for data acquisition are accomplished with a Daytronic 9000 System equipped with a 9635 computer interface module. The control and data acquisition system hardware design and specification as well as the associated computer software were developed by the Commercial Roofing Laboratory.

The sample grips are relatively simple, single unit mechanical grips. The faces of the gripping surfaces are transversely serrated to enhance gripping. The grips are tightened at the beginning of the test and need no further adjustment. For EPDM and PVC materials, additional separate pieces of the test material were used to pad both faces of the grips at both temperatures. Modified bitumen materials can be tested in the grips without additional padding at room temperature, but required padding to prevent grip failures at -18 degrees C (0 degrees F). PVC was found to function well to pad the modified bitumen samples. The grips are single unit rather than the multiple pivot grips often used in biaxial testing[3]. The single unit grips are simpler and allow reduced time for sample preparation and mounting. If sufficiently uniform loading could not be achieved with these grips, the multiple pivot type would have been the alternative. However, all gripping was judged to be satisfactory with the procedure described above and the single unit grips.

The BRST has been used primarily to verify computer analyses of roofing systems under changing thermal loading. In these cases, the motion of any portion of the test sample is quite small, and large motions of the sample grips are not required, since the primary quantities being measured are loads rather than displacements. However, the current testing of biaxial failure strengths and elongations requires grip motions similar to a standard tensile test. When testing biaxial properties, the four lengths of material from the grips to the biaxial test area are in uniaxial tension and will behave much as the material would in a standard tensile test. Therefore, in order to accommodate a wide range of elongations for various roofing materials, a much smaller sample size was chosen so that sufficiently high elongations in the uniaxial portion of the sample could be attained with the grip travel limited to 23 cm. Only the central area of the specimen is in a state of biaxial stress. This area will exhibit stress-strain behavior which is considerably different from the uniaxial regions and will produce the failure loads at smaller elongations than seen in uniaxial tests.

SAMPLE CONFIGURATION AND PREPARATION

The sample geometry chosen for testing is the 'slitted cross shaped specimen' which has been found to yield the best results for coated fabric samples, (Reinhardt[4]). All samples were cut using a template and a utility knife, with the exception of the rounded corners at the intersection of the axes on the biaxial samples. These corners were punched in a 4.8 mm (3/16 in) radius. Two different length samples were used in the testing. For materials which were expected to exhibit high elongation, the overall sample length on both axes was chosen to be 47 cm (18.5 in). This provided for 10.2 cm (4 in) of material in each grip and a gage length of 27 cm (10.5 in) on both axes. This geometry combined with the available grip travel will allow maximum elongations to failure of 175 percent biaxially. For lower elongation materials, the total sample length was chosen to be 67 cm (26.5 in) on each axis, with a gage length of 47 cm (18.5 in).

Each of the uniaxial tabs which connect the biaxial area to the grips contain longitudinal slits. The function of the slits is to prevent the loading along one axis from dissipating into the width of the perpendicular axis. Just as the load per unit width would be reduced in the central section of the uniaxial specimen illustrated in Figure 1a, so would the load per unit width be reduced in the center of a biaxial sample if the slits were not included. The slits prevent the uniaxial loading tabs of the sample from supporting transverse load. When the slits are present, the load per unit width remains constant on both axes even through the biaxial central section (Figure 1b). For the lower elongation materials, each sample axis end contained three slits beginning at the biaxial area and extending toward the grips (Figure 1c), which divide each uniaxial segment into four tabs of equal width. Samples for higher elongation materials were prepared with seven slits, three primary slits as with the low elongation materials and four secondary slits. The secondary slits divide each of four tabs created by the primary slits into two tabs, resulting in eight equal width tabs per side of the biaxial area. The sample slits were hand cut using a straight edge and a utility knife.

All slit tips are blunted by punching holes in the material at the slit ends at the biaxial area, to prevent any slit tip from functioning as a crack tip and propagating the crack into the biaxial area. All holes were 2 mm (.08 in) in diameter and were punched into the materials using a common leather punch. The three major slits terminate at holes adjacent to the biaxial area. The secondary slits terminate at holes which are staggered away from the biaxial area by one hole diameter.

The configuration of the test samples must be chosen carefully. Generally, the minimum number of slits that perform satisfactorily should be used. Greater numbers of slits create a larger number of narrower tabs. The narrower tabs are more likely to isolate local weak spots in the test material than are wider tabs. Further, there are more opportunities to sever longitudinal reinforcing strands which are not perfectly aligned with the sample axes on materials with scrim or woven reinforcements. In both cases, the result is a weakened tab which is more susceptible to premature tab failure. However, more slits provide more even stress distribution at the edges of the biaxial area and lower the stress concentrations at slit tips. In practice, the modified bitumen products tested better with samples containing only three slits, but the EPDM and PVC samples fared better on samples with seven slits with staggered tips as described above. The holes should be large enough to reduce the stress concentration at the slit tip to an acceptable level, but should be kept as small as possible so that the tabs are not significantly weakened by removal of material. The selection of slits and hole sizes required judgement and some preliminary testing, resulting in the sample configuration described here.

FAILURE

Two indicators were considered in determining whether sample failures were valid, actually representing the biaxial failure limit of the material, or premature, failing as a consequence of inadequate sample design or fabrication. The foremost consideration was that the failure must be associated with the biaxial area. Uniaxial tab failures were not considered to be valid sample failures. Local irregularities in material strength along a narrow tab can lead to an early failure of the tab before the biaxial area is stressed to its failure limit. The type of failures

anticipated with the plane biaxial sample are expected to
initiate at the edges of the biaxial section. As
fundamental principles of stress analysis would indicate,
and as is illustrated photoelastically by J. C. Radon, et
al[3], the slits in the sample which insure uniform biaxial
loading also introduce a degree of stress concentration at
the slit ends whether they are blunted or not. Therefore,
valid sample failures can be expected to initiate at the
slit tips since this will form the weakest link. However,
if the material is at its biaxial failure limit, the failure
should propagate readily into the biaxial area though it was
initiated on the edge. As an example, consider the
difference between uniaxially and biaxially stressed EPDM.
EPDM samples stressed uniaxially to near their uniaxial
failure limit will not readily propagate an induced
transverse slit across the sample width. In sharp contrast,
for biaxial EPDM samples stressed to near their failure
limit, the initiation of the failure and its propagation
through the biaxial area occurs so quickly that it appears
to be instantaneous. The PVC and modified bitumen materials
failed more gradually than EPDM, but both materials
exhibited tears or cracks which propagated into the biaxial
area at failure.

The final principle used in determining failure was
whether the material had sustained permanent damage. Failure
in some of the materials was not instantaneous but was a
process which occurred over a short time interval. Tears in
the material propagated into the biaxial area or material
reinforcement components failed. In both cases, the sample
load began an irreversible drop. At this point, the sample
has failed. It will no longer perform as the virgin
material and would be unsuitable for service as a roofing
material. Once irreversible damage which decreases the
ability of the material to withstand loading has occurred,
the sample has failed.

There is no consensus on the relation of biaxial to
uniaxial failure strengths for flexible membrane materials.
R. Mott, et al[4], reported biaxial strengths which were
considerably reduced from uniaxial values using cylinder
tests on TFE coated woven glass fabrics. However,
Reinhardt[1] demonstrated progressive increases in the ratio
of biaxial to uniaxial failure strengths through progressive
improvements in sample configuration and gripping procedure.
The final improvement resulted in a biaxial failure strength
equal to the uniaxial failure strength for a coated woven
fabric material. The actual ratio of biaxial to uniaxial
strengths may well be material dependent. Whether or not
the biaxial failure load is equal to the uniaxial load or
reduced, in the current testing biaxial strengths were
required to be a significant fraction of the uniaxial values
for the test to be considered valid.

The biaxial elongations are expected to be significantly
lower, especially for the high elongation materials, owing
to the effect of the off-axis loads and Poisson's ratio.

For homogeneous isotropic materials, the basic equations of elasticity predict a biaxial elongation at failure which is reduced to a value of (1-p) times the uniaxial value, where p represents Poisson's ratio, if the biaxial and uniaxial failure stresses are the same.   However, many roofing materials are non-homogeneous anisotropic composites.   In these cases, the basic equations of elasticity are inadequate to fully describe the materials.   Further, the question of how the biaxial failure stress compares to the uniaxial must be considered.   In any case, the biaxial elongation values are expected to be considerably reduced from the uniaxial values, especially for the high elongation materials.

TEST CONDITIONS

Tests were performed under conditions typical for roofing materials.   Two test temperatures were used, room temperature and -18 degrees C (0 degrees F).   Materials with uniaxial strains greater than 75 percent were tested at a strain rate of .1 cm/cm per minute, and materials with lower elongations were tested at a strain rate of .01 cm/cm per minute.   Since all four grips were active, the grip speed of each grip can be found by multiplying one half the sample length by the strain rate.   However, realizing that the biaxial area will strain less readily than the attached uniaxial tabs, it should be noted that the biaxial area will be strained at a lower rate than input to the grips while the uniaxial sections are straining at a higher rate.   In addition, these rates may not always be constant nor proportional.   In the case of the PVC material, the biaxial area elongates only slightly while the uniaxial tabs undergo high elongation while they permanently deform to a necked down geometry.   Once the necking process is complete, the biaxial area rate of elongation becomes closer to the rate of the uniaxial tabs.   The EPDM and modified bitumen materials exhibit more proportionality between the biaxial and uniaxial elongations.

RESULTS

Graphical results of typical load-strain data for the five materials tested are presented in Figures 2 through 13. The load-strain graphs contain both the biaxial and uniaxial data for a representative individual sample for each material.   As these graphs depict results for individual samples, they will not exactly match the values in Tables 1 and 2, which are averages of multiple samples.   Only the cross machine behavior is shown on the graphs when it is representative of testing in both directions.   There was very little difference between the machine(MD) and cross

machine(CMD) performance for both the EPDM and PVC
materials; they are essentially orthotropic. Therefore, only
CMD results are given. However, both the glass/polyester
composite modified bitumen and the polyester reinforced
modified bitumen show significant differences between the
machine and cross machine behavior.   For these materials,
both the MD and CMD results are shown.

The most obvious feature of these graphs is that the
biaxial elongations are significantly lower than the
elongations of the uniaxial tabs.   In Figures 6 and 9 the
biaxial elongation is seen to be nearly exactly one half of
the uniaxial tab elongation.   This would indicate that the
rubber material was behaving ideally with a Poisson's ratio
of .5, the appropriate value for an incompressible material.
Figures 7 and 8 show similar results but are not as precise
in the relationship.   The PVC membranes tested also
demonstrate a similar reduction in biaxial failure
elongation as shown in Figures 10 and 11.   However, the
modified bitumens were not so well behaved. The CMD results
for the polyester reinforced product show a loss well in
excess of 50 percent, while the MD results indicate a loss
of 50 percent or less (Figures 4 and 6).   Finally, the
glass/polyester composite reinforced product indicated
biaxial strains that were equal to or even greater than the
uniaxial failure elongation (Figures 2 and 3).   It should be
noted that the elongation values for the glass/polyester
modified bitumen may be subject to greater percentage errors
than the other materials.   It is thought that a slight
amount of initial sag on these samples may have contributed
to the unusually high biaxial values.   In measuring small
strains, small absolute errors can cause large percentage
errors.   While the exact amount of the error could not be
precisely determined, some approximate corrections were
attempted and resulted in the second set of values in Table
2 for the biaxial strain.   This yielded values of 2.0 and
2.4 for the biaxial and uniaxial values respectively.

The second item of note is that the uniaxial tab
elongation is also smaller than the reference uniaxial
elongation for the materials; see Tables 1 and 2. ✒ This
reduction in elongation to failure is quite pronounced in
all of the high elongation materials, but was not
necessarily expected. If the biaxial failure strength of
the material was equal to the uniaxial, there should be no
reduction of the uniaxial tab elongation at failure of the
biaxial section. As a result, the biaxial failure strength
is expected to be lower than the uniaxial value for these
cases, which is indicated by the data. Comparisons of all
the materials can be made with the data in Tables 1 and 2.

Tables 1 and 2 contain the average data for each
material. Three replicates were tested for each material at
room temperature, and two were tested at -18 degrees C.
Both strains are given for the biaxial sample; the strain of
the central biaxial area and the strain of the uniaxial

sample tabs.   A reference failure strain obtained from the testing of small laboratory samples is also listed.

Figures 14 through 19 are included to provide some insight into how a typical test would appear.   All of these tests were performed at room temperature.   A three picture sequence to failure is given for 60 mil EPDM in Figures 14, 15, and 16.   A PVC and a polyester modified bitumen failure are depicted in Figures 17 and 18 respectively. Figure 19 shows the failure of a glass/polyester modified bitumen at the intersection of the lower left tab with the biaxial area. Figure 20 shows the test area with the environmental chamber tilted back away from the biaxial test area.

Table 1 contains two sets of data for 60 mil EPDM. The EPDM material tended to be so consistent in testing that these results were unusual.   Four samples were tested resulting in two pairs of results.   Since each pair was so consistent and distinct, both are listed.   One sample of the pair with lower failure values was observed to have an air bubble in the interior of the material at the failure surface.   However, no defect was found with the second sample.

CONCLUSIONS

1. Biaxial testing of roofing materials is feasible and should be pursued as a prelude to the establishment of performance based specifications.

2. Biaxial elongations to failure are well below uniaxial failure elongations for several common roofing materials.

3. Further studies are needed to quantify the reduction of failure strength of biaxial samples for specific materials. Future studies should also include the effects of aging on the materials.

4. The service environment for roofing materials is biaxial, and it has been shown that significant differences exist between biaxial and uniaxial properties.   Performance specifications for roofing systems would be most appropriate if developed in light of biaxial as well as uniaxial data.

Table 1 --   Room Temperature Results

Material	Sample Load (Newtons)		Sample Biaxial Strain %		Sample Uniaxial Strain %		Reference Uniaxial Load/Strain
	MD	CMD	MD	CMD	MD	CMD	CMD
Mod Bitumen glass/poly	1700	2500	2.7	2.8	2.1	2.1	2970 /3.4[a]
Mod Bitumen polyester	1530	800	15	9.5	32	37	1050 /49[a]
EPDM 45 mil	470	490	77	100	163	196	1120 /370[b]
EPDM 60 mil	640/420	540/430	85	80/60	159	180/128	1690 /350[b]
PVC 48 mil	765	740	32	40	105	150	1370 /250[c]

Table 2 --   (-18) Degree C Results

Material	Sample Load (Newtons)		Sample Biaxial Strain %		Sample Uniaxial Strain %		Reference Uniaxial Load/Strain
	MD	CMD	MD	CMD	MD	CMD	CMD
Mod Bitumen glass/poly	5140	4520	3.5/2.0	3.5/2.0	2.4	2.4	5600 /2.9[a]
Mod Bitumen polyester	2290	1530	11.8	8.7	16.6	16.6	2240 /34.2[a]
EPDM 45 mil	820	870	92	96	153	150	1780 /302[b]
EPDM 60 mil	960	840	85	75	141	146	1670 /246[b]
PVC 48 mil	1570	1540	15.1	11.8	46	48	... /...

[a] O.C.F. testing ASTM D2523
[b] O.C.F. testing by ASTM D412; D412 strain rate is 6.7/min
[c] Literature values by ASTM D638

[1] Radon, J. C., Leevers, P. S., and Culver, L. E., "A Simple Testing Technique for Fracture under Biaxial Stresses," Experimental Mechanics, June 1977, pp.228-232.
[2] Mott, R., Huber, G., and Leewood, A.,"Characterizing Permanent Fabric Roof Materials," Journal of Testing and Evaluation, Jan 1985, pp 12-18.
[3] Testa,R. B., and Boctor, W., "A Biaxial Stress Transducer for Fabrics," Experimental Mechanics, March 1984, pp.33-40.
[4] Reinholdt, H. W., "On the Biaxial Testing and Strength of Coated Fabrics," Experimental Mechanics, February 1976, pp 71-74.

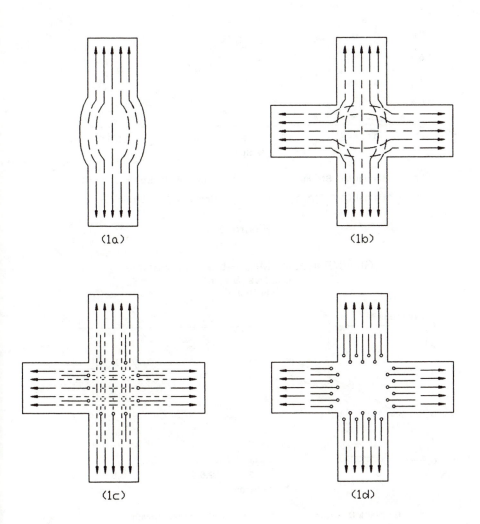

(1a)   (1b)

(1c)   (1d)

Figure 1

Glass/Polyester Mod Bit Biaxial Sample
Load vs. Strain
Room Temperature

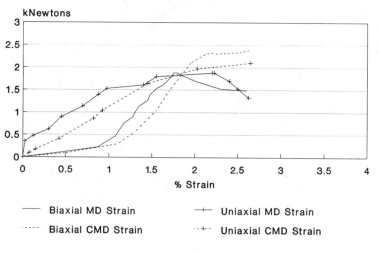

— Biaxial MD Strain   —+— Uniaxial MD Strain

···· Biaxial CMD Strain   ··+·· Uniaxial CMD Strain

Figure 2

Glass/Polyester Mod Bit Biaxial Sample
Load vs. Strain
-18 Deg. C

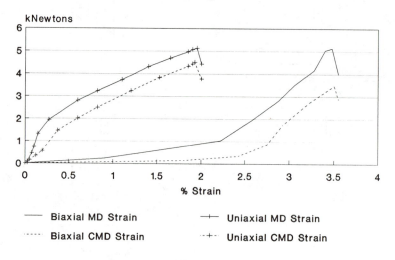

— Biaxial MD Strain   —+— Uniaxial MD Strain

···· Biaxial CMD Strain   ··+·· Uniaxial CMD Strain

Figure 3

Polyester Mod Bit Biaxial Sample
Load vs. Strain
Room Temperature

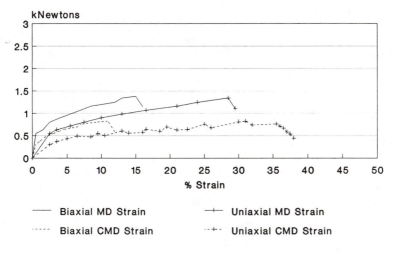

—— Biaxial MD Strain	—+— Uniaxial MD Strain
----- Biaxial CMD Strain	--+-- Uniaxial CMD Strain

Figure 4

Polyester Mod Bit Biaxial Sample
Load vs. Strain
-18 Deg. C

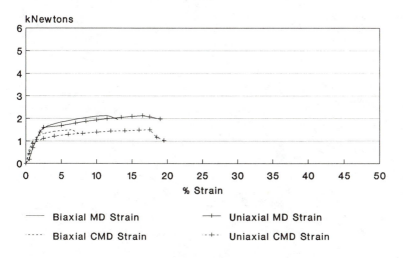

—— Biaxial MD Strain	—+— Uniaxial MD Strain
----- Biaxial CMD Strain	--+-- Uniaxial CMD Strain

Figure 5

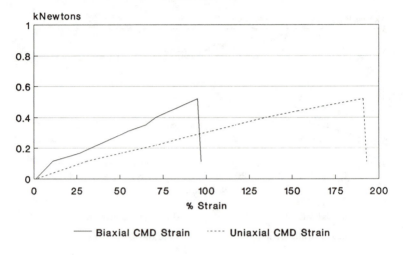

EPDM 45 Mil Biaxial Sample
Load vs. Strain
Room Temperature

Figure 6

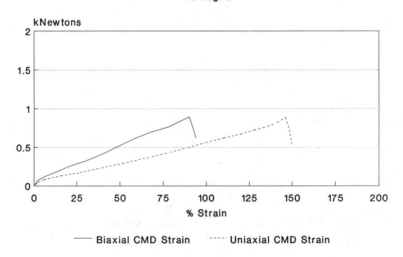

EPDM 45 Mil Biaxial Sample
Load vs. Strain
-18 Deg. C

Figure 7

EPDM 60 Mil Biaxial Sample
Load vs. Strain
Room Temperature

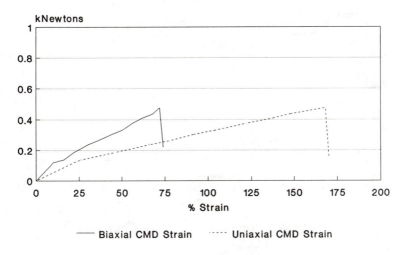

Figure 8

EPDM 60 Mil Biaxial Sample
Load vs. Strain
-18 Deg. C

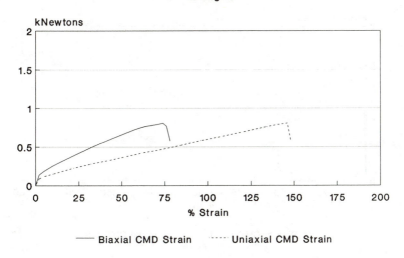

Figure 9

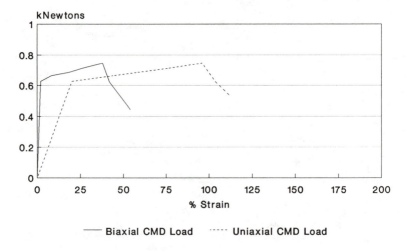

PVC Biaxial Sample
Load vs. Strain
Room Temperature

Figure 10

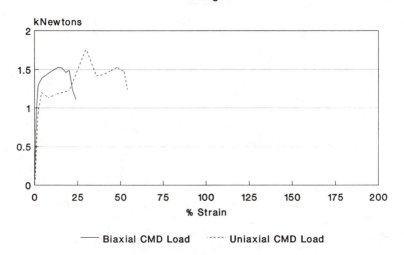

PVC Biaxial Sample
Load vs. Strain
-18 Deg. C

Figure 11

Biaxial Product Comparison
CMD Load vs. Strain
Room Temperature

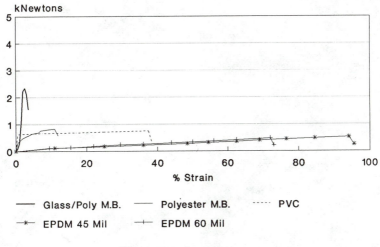

Figure 12

Biaxial Product Comparison
CMD Load vs. Strain
-18 Deg. C

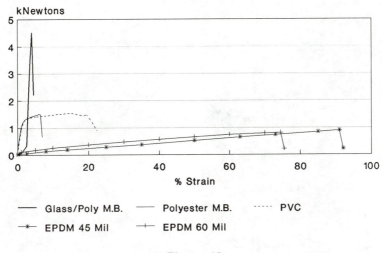

Figure 13

Figure 16 - EPDM 60 mil failure

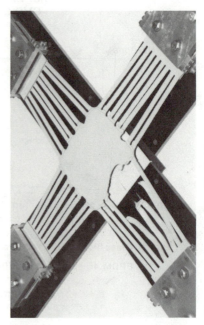

Figure 17 - PVC 48 mil failure

Figure 14 - EPDM 60 mil before test

Figure 15 - EPDM 60 mil mid test

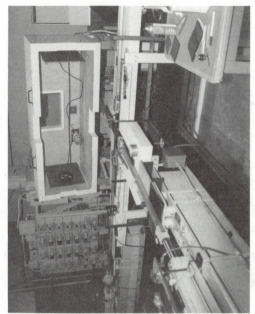

Figure 20 - Biaxial Roof System Tester

Figure 18 - Polyester Mod Bit Failure

Figure 19 - Glass/Poly Mod Bit Failure

Mark Easter

FINITE ELEMENT ANALYSIS OF ROOFING SYSTEMS

REFERENCE: Easter, Mark R., "Finite Element Analysis
of Roofing Systems", Roofing Research and Standards
Development: 2nd Volume, ASTM STP 1088, Thomas J.
Wallace and Walter J. Rossiter, Eds., American Society
for Testing and Materials, Philadelphia, 1990.

ABSTRACT: Finite Element Analysis has been used for
almost 30 years by areospace and automobile industries to
design parts and systems. Finite Element Analysis, or
FEA breaks a part up into small elements in order to find
the stresses on the whole. This paper describes how the
method was applied to single ply roofing. Standard wind
test assemblies were analyzed for verification and then a
corner area of a typical roof was analyzed.

KEYWORDS: Finite Element Analysis, single ply,
modulus, reinforcement, stress, strain

Wind resistance of roofing systems is currently designed
around either negative or positive pressurized wind uplift tests.
Finite Element Analysis or FEA is another tool that can be used to
design roofing systems, for wind resistance. FEA is a
mathematical analysis model of an assembly. FEA simplifies an
analysis problem by breaking it into small parts and adding up the
results.

FEA is done by dividing the whole part or assembly model to be
analyzed into many smaller volumes or areas. The effect is a
checkerboard pattern or grid drawn on a computer screen by a
preprocessing program. The intersections of the dividing lines are

Mr. Mark R. Easter is a Systems Engineer at Firestone Building
Products a division of Bridgestone/Firestone Inc., 525
Congressional Blvd., Carmel, Indiana 46032.

called nodes. The forces acting on objects are then broken down and theoretically applied at each node. The node and its surrounding area exert force to the adjacent nodes and areas.

Equations that describe the forces each area or element exerts on the other are filled in by the computer program. These equations are then solved simultaneously, since the force or node is dependent on many others.

This method has been known since the 1940's but wasn't used until the 1960's when computers were able to solve these complex problems. It wasn't until the late 1970's that computers in general use became powerful enough for wide spread use of FEA. Today workstations costing as little as $10,000 can solve these complex problems.

The FEA method was originally designed for aircraft and aerospace applications. FEA allowed parts to be designed with minimum size and weight, resulting in lighter planes. Automakers now use this method to eliminate model making and testing to speed up product development.

Roofing is also a good application for FEA. Some roofing materials have elongation that make conventional design or analysis methods applied to metal parts impossible. Non reinforced single plys have enough elongation that they can not be analyzed by rubber design guidelines. Hence, the advantage of FEA.

FEA can be used to analyze reinforced and non reinforced roofing systems. Traditional tests like Factory Mutual (FM) 5' x 9' wind uplift tests are fairly well understood. Force transducers can be placed in anchoring locations and fastener loads measured. Elongations of membranes along battens show forces on the battens. FEA analysis of 5' x 9' tests have shown results very close to actual tests. FEA is very useful for studying larger testing table sizes. On larger than 1.525m x 2.743m (5' x 9') test sizes there is much less edge effect, that is, the membrane attachment fasteners hold more load then the edges of the test fixture. Analysis of the 5' x 9' test was the first step in finding the applicability of FEA to Roofing Systems. It provided an easy way to verify initial results.

The 5' x 9' test was analyzed with a single ply EPDM batten in the seam system. Batten strips are 2.134 meters (7 feet) apart and are fastened every 0.305 meter (1 foot) in the test assembly. Batten strips are placed parallel to the 1.524m (5 feet) dimension.

MATERIAL PROPERTIES

The first step in FEA analysis is to determine the material

properties. EPDM is a relatively elastic material that exhibits some elastic creep and a low modulus. A very slow pull rate was used to find the Youngs modulus of the material which was about 2.689 MPa (390 psi). The slow rate simulates a wind uplift test that takes 6 minutes or longer. Slight non linearity at very high elongations resulted in a modulus of 2.586 MPa (375 psi) being used in the analysis. See Figure 1.

BUILDING THE FEA GRID

The FEA grid is made up of triangles, squares or polygons that fill in the geometry of the part. A membrane will have one layer of elements, a thicker part might have many layers. A membrane shell element which resists force only in plane was chosen as most representative of the roofing membrane. This type of element doesn't model thickness. The shape of the element is a triangle and it was of a linear order with 3 nodes and three degrees of freedom per node. A element size of 7.62 centimeters (3") was used for this analysis. Larger size elements could cause the FEA equations to not converge, that means the FEA equations become unsolvable. Smaller sizes are a better representation of a homogeneous material but take much more computer time to solve. A trial analysis using solid brick elements was performed but numerical difficulties were noted and the attempt was dropped. The solid elements added little to the analysis other than the membrane thinning which occurs during the deformation of the membrane. The cost of using the solid elements is also much higher than the membrane shell elements. Thus for most work it would appear that the effort required to use the solid elements is not worth the cost.

Patran (PDA Engineering) was the program used to build the FEA models. SAFEM (Applied Mechanics Inc.) was used for the Finite Element Analysis. A critical feature of the finite element program used is its ability to model large geometry changes. The amount of deformation that occurs during the loading of the membranes as well as the large strains encountered require that a nonlinear geometry with both large strain and large deformation theory be used. Simple linear elastic FEA programs will not provide adequate results.

The only real limitation of the shell elements is that they do not thin as they are stretched thus introducing some error in stiffness as the membranes are grossly deformed. This was noted and some correction was approximated by reducing the material stiffness to allow for a softer response. Future analyses could be enhanced by modifying the membrane shell elements to allow thickness change during deformation. For this initial study it was not felt to be cost effective and the main desire here is to mention this as an aid to interpretation of the results.

Any node in the grid can be restrained in one, two or all three degrees of freedom. The edges of the testing table, like the edges of a roof are fixed so they are assigned 0 degrees of freedom in the analysis.

Battens were approximated by fixing a line of nodes with zero width. The force required to restrain the line or batten can be found at every node point. Observations of experiments on tables indicated that there is some error involved as the batten strips are flexible and actually lift off the table between the screws fastening them to the deck. Only vertical forces on the Batten Strip itself (screw pullout forces) can be determined with this modeling technique.

RESULTS

One benefit of advanced FEA programs is the graphical results they display. These are shown in Figures 2 thru 6. They show maximum fastener loads of 1.245 kn (280 lbs.) and strains of 90% for 5.746 kPa (120 psf) pressures. Ballooning heights of 1.829 m (6 feet) are shown (Figure 2). Figure 3 shows stress in the long direction against the Batten Bar. Stresses in the short direction against the long edges of the table are shown in Figure 4. Figures 3 and 4 show that the stress at the edges of the table are higher than the stresses at the batten strips. This is called edge effect. Since one fastener holds 0.541 square meter (7 square feet) of membrane a loading of 7 sf x 120 lb/sf) or 3.736 kn (840 lbs.) is possible.

VERIFICATION

Strain in actual 5' x 9' tests can be measured by marking lines on the test specimens and measuring their elongation. It is not safe to measure these at pressures above 90 psf. However, actual measurements show 60 (max) percent elongation at 4.309 kPa (90 psf) vs. 64 percent elongation according to FEA analysis, a very close correlation.

Fastener loads have not been measured for batten systems. Analysis of 0.914 m (3') o.c. a spot attached system test showed 2.224 kn (500 lbs.) actual load and the FEA analysis showed 2.202 kn (495 lbs) for 4.309 kPa (90 psf) loading in the 5' x 9' test.

12 x 24 TEST

The FM 12' x 24' wind uplift test was the second Finite Element analysis performed. Not many tests have been run to date and test results vary widely. This test was analyzed by taking advantage of symmetry about the central batten as shown in Figure 7. The results show much less edge effect in 12 x 24 testing. Figures 8 and 9 show results. Bubble heights in these tests can be over 2.438 m (8 feet). Results are shown for 15.240 cm (6") o.c. fastening and can be multiplied by 2 for 30.480 cm (12") o.c. fastening. 4.309 kPa (90 psf) is the highest pressure analyzed because of the high loading and severity of this test.

One interesting problem which developed during the analyses was due to the large ballooning which occurs during higher inflation pressures, the multiple bubbles on a table or roof will contact each other. In straight forward analyses, the numeric computations actually allow these bubbles to pass through each other. This of course does not represent the real test and to solve this problem contact type elements were employed to restrain the bubbles from penetrating one another. The interaction of the bubbles tends to decrease the fastener force in general so its neglect in certain cases adds to the margin of safety. Sufficient use of the contact elements was made to demonstrate their effectiveness and their use mainly adds to the time required to construct a model, not the computer time to analyze it.

Thermoplastic and reinforced membrane have higher moduli (are stiffer) than non reinforced EPDM. The effect of other moduli was also investigated. Moduli of 2 to 5 times the "standard membrane" could be possible with non reinforced membranes. Moduli of 20 to 50 times standard represent reinforced membrane. The results show that reinforced membranes actually exert more force on attachments. This makes sense because lower modulus would better distribute force. Non reinforced material transmits more load to the edges and flattens out the peak load on the fasteners in the center of the batten. Elongation and bubble heights are of course much lower with reinforced membrane. See Figure 9.

A SIMPLER MATHEMATICAL MODEL

A simple mathematical model was used as a check on the results. Superposition was used to find the load on the center fastener of the center batten. Rubber was modeled as a spring system with distance to the batten and sides being the initial lengths of the spring. A load of 266.880 and 400.320 n (60 and 90 lbs.)

was applied to the three springs in parallel. The change in length of the three "springs" was found and then the load carried by the "spring" attached to the batten could be found by its elongation, (F = kx). See Figure 10.

This is an approximation; not exact because the membrane is a continuous sheet not just .305 m (1 foot) wide strips. It does give an example of how much load is still carried by the edge - especially by wider sheets. See Figure 10.

LIMITATIONS

FEA analysis of 5' x 9' and 12' x 24' testing can only predict strain and loading on membrane and fasteners. It cannot account for real world testing factors. Continued load on fasteners (and decking) can cause pullouts lower than these found in pullout tests. Loading caused by tests such as the 5' x 9' pull on the fastener at 30° to 60° angles. This can also cause lower pullout. Adhesives and seam tapes can also fail at lower constant loads. FEA could eliminate the need for large scale wind uplift testing by predicting these large scale loads. A 5' x 9' test could screen the testing factors such as adhesive characteristics.

A roof corner system was analyzed with the proven FEA methods. Southern Building Code Congress (SBCCI) corner and perimeter pressure coefficients were used. Figure 11 shows how the model was laid out. Symmetry was not used in this analysis. The 36' corner area includes 3 - 1.524 m (5') perimeters and 3 - 2.134 m (7') field sheets. The maximum velocity uplift pressure of -1.197 kPa (-25 psf) represents up to a 49.170 m/s (110 mph) wind. The small rectangular areas with a 1.7 coefficient were added for a consistent computer analysis between corner and perimeter areas. This analysis showed similar loading in 1.524 m (5') perimeters and 2.134 m (7') field rows of battens. It also showed stress concentrations in corners that might not have been expected on the first examination. Actual wind loading formulas such as those by Phalen[3] could be substituted for the coefficients for a more precise analysis.

APPLICATIONS FOR FUTURE FEA ANALYSIS

Other possible areas for FEA analysis are insulations. An insulation in a fully adhered system serves as a structural member. It is usually a composite of facers and a center of foam plastic. FEA could find the internal and surface stress in the foam, especially around fasteners and plates. Changes in the insulation could result in less fasteners being used. Optimal fastener spacings could also be found.

Roofing components such as spot attachment anchors could also be analyzed and optimized structurally. This would be more in line with traditional automotive and aerospace uses of FEA for component design.

As roofing systems become more sophisticated and cost competitive FEA will be a valuable tool for design of wind resistant roofing systems.

REFERENCES

(1)   Dubensky, R. G., What Every Engineer Should Know About Finite Element Analysis Methods, SAE 861294, SAE Warrendale, PA. 15096, 1986.
(2)   Lightner, John J., Membrane Roofing System Analysis Via FEA, Bridgestone/Firestone Inc., Akron, Ohio 44317, 1988.
(3)   Phalen, T. E., Design Considerations For Lightweight Concrete Ballast In Loose Laid Roof Systems, Lincoln College, Northwestern University, Boston, MA

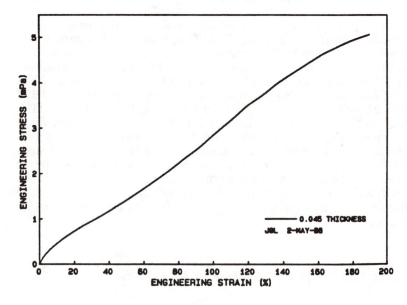

Figure 1. EPDM Roofing Membrane Stress-Strain Reponse

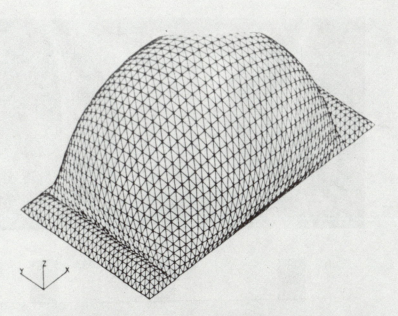

Figure 2. Deformed Shape at 120 psf of 5 x 9 Table with Batten

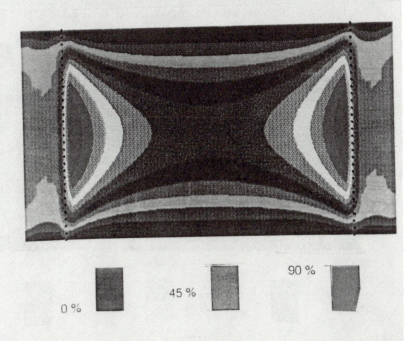

Figure 3. Stresses in the Long Side Direction at 120 psf for 5 x 9 Table

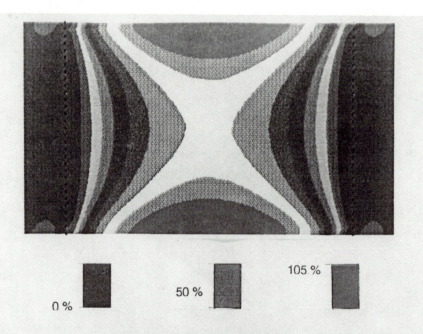

Figure 4. Strains in the Short Side Direction at 5.744 kPa (120 psf)
for 5 x 9 Table

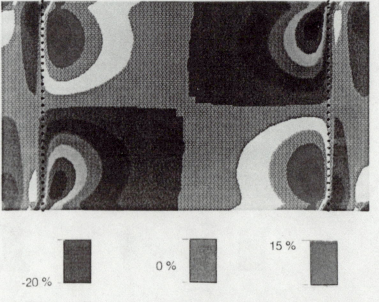

Figure 5. Shear Stresses at 5.744 kPa (120 psf) for 5 x 9 Table

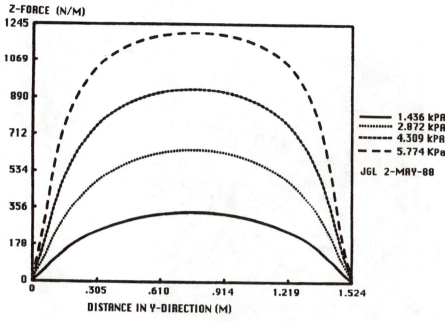

Figure 6. Vertical Batten Forces for 5 x 9 Table

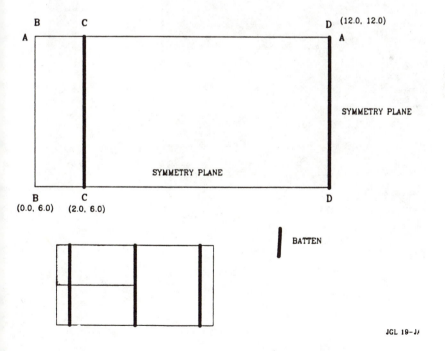

Figure 7. Layout of 12 x 24 Test With 3.04m Spacing

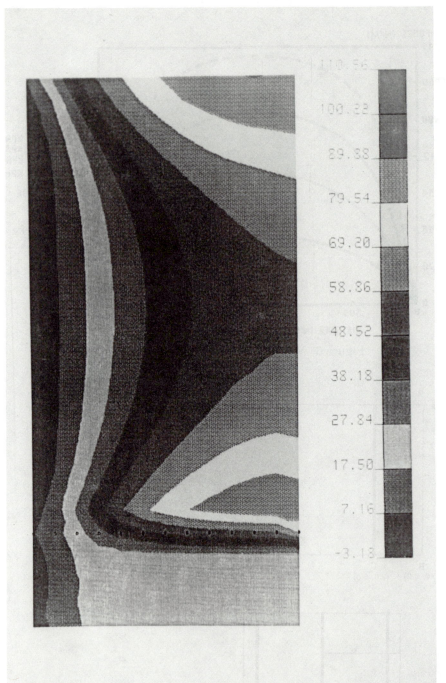

Figure 8. 12 x 24 Table With 3.04m (10 ft) Batten 2.872 kPa (60 psf)
Long Side Stresses

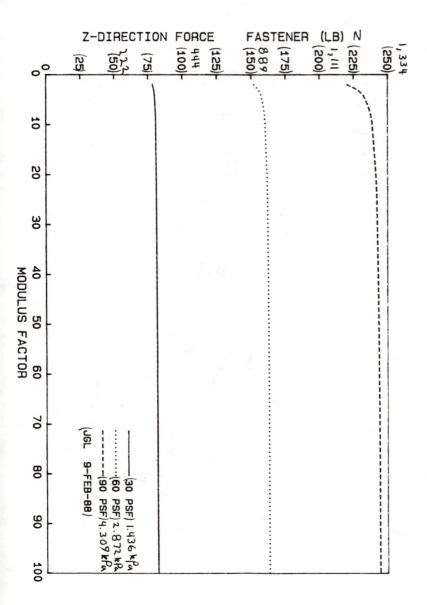

Figure 9. 12 x 24 Table with 3.048m (10 ft) Batten, Edge Batten Peak
Vertical Force versus Modulus

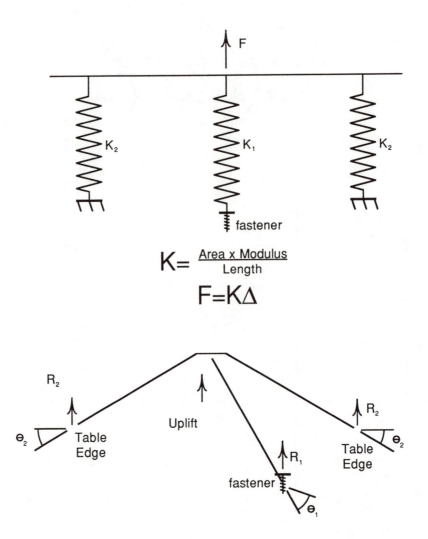

Figure 10. Spring Model

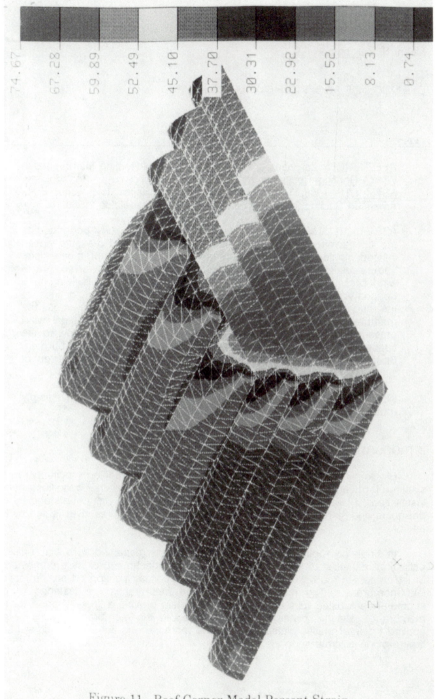

Figure 11. Roof Corner Model Percent Strain

Mehdi S. Zarghamee

DYNAMICS OF ROOFING MEMBRANES IN WIND

REFERENCE: Zarghamee, M. S.[*], "Dynamics of Roofing Membranes in Wind," Roofing Research and Standards Development: 2nd Volume, ASTM STP 1088, Thomas T. Wallace and Walter J. Rossiter Eds., American Society for Testing and Materials, Philadelphia, 1990.

ABSTRACT: This paper presents the results of a study performed to predict the effect of the dynamic component of wind pressure on an inflated single-ply roofing membrane and its attachment mechanism. The calculation is performed based on the measured power spectra of wind pressure at points on the roof of a square plan building in a boundary layer wind tunnel. The transient response factor, defined as the ratio of the fluctuating component of pressure difference across the membrane thickness to the fluctuating component of external wind pressure, is shown to depend strongly on the ease of air flow to the space underneath the membrane and to the size of the building. Specific recommendations are tentatively proposed for the design of single-ply roofing systems against wind uplift.

KEYWORDS: Dynamics, infiltration, membrane, roof, roofing, single-ply roofing, wind, wind loads.

INTRODUCTION

Single-ply roofing systems are selected over the traditional built-up roofing systems for their lower cost and their lower sensitivity to weather conditions during installation. However, their behavior in wind has not been well understood by the designers and manufacturers; this has resulted in many failures investigated by the author.

In single-ply roofing systems, the membrane is restrained from wind uplift by ballasting, adhering or heat-bonding, mechanical attachment, or by a combination of the three. In the ballasted system, gravel or pavers are applied on the loosely laid membrane. The mechanically attached membranes are fastened to the structural deck using bars or washer plates. Bars provide a linear support for the membrane. Washer plates are either applied to the membrane directly or are applied to larger plastic plates to which the membrane is adhered; in either case, they provide discrete supports.

[*] Principal, Simpson Gumpertz & Heger Inc., 297 Broadway, Arlington, MA 02174

In typical modes of failure of single-ply roofing systems, the membrane becomes inflated and billows. The failure occurs either in the system of restraint and attachment or in the membrane itself.

Adhered systems usually contain zones of poor bond that inflate in wind. The restraint against uplift is provided by the peel strength of the membrane from its substrate. The inflated membrane zone will grow if the peel strength of the adhered membrane is less than the imparted wind load. The membrane of mechanically attached systems are normally inflated in wind. The restraint against uplift and growth of the inflated membrane zone is provided by mechanical fasteners and, to a lesser extent, by membrane stiffness. Ballasted systems also fail by the inflation of the membrane after the ballast has been dislodged (see Kind and Wardlaw [1 and 2]). In such a case, the pile of the dislodged ballast at the perimeter of the inflated membrane zone provides the restraint against uplift and its growth.

In all these cases, the roofing membrane is not subjected to the full wind pressure unless free air can flow to the space underneath the membrane.

This paper introduces some of the factors that influence the dynamic wind loads subjected to roofing membranes and to their systems of restraint and attachment. Such loads are determined based on the power spectra of wind pressure measured on the roof of a square-plan, flat roof building in a wind tunnel.

THEORETICAL BASIS FOR CALCULATING RESPONSE OF AN INFLATED MEMBRANE TO WIND

The notation used below is defined in Fig. 1.

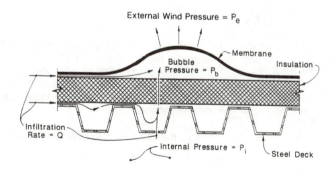

FIG. 1 – Behavior of inflated membrane – notation

When an inflated membrane is subjected to an external time-varying pressure, P_e (as shown in Fig. 1), while the internal pressure, P_i, is constant, conservation of mass requires that the following relationship be held between the change of the pressure inside the bubble, P_b, and the change of the bubble volume, V_b [3]:

$$\frac{d\left(V_b\, P_b\right)}{d\,t} = Q(t)\, P_i \tag{1}$$

where Q is the infiltration rate at time t.

The rate of air infiltration into the bubble at any time, t, is governed by the expression (see ASHRAE Handbook [4])

$$Q(t) = C\, A\, [P_i - P_b(t)]^{0.65} \qquad \text{for } P_i \geq P_b(t)$$

$$= -C\, A\, [P_b(t) - P_i]^{0.65} \qquad \text{for } P_i < P_b(t) \tag{2}$$

No measurement of air infiltration to the space underneath a roofing membrane is reported in literature; however, Tamura and Shaw [5] have measured the rate of air infiltration through walls of tall buildings and found it to be in the range of 0.61 to 2.44 liter/sec·m^2 (0.12 to 0.48 cfm/ft^2) under a pressure head of 75 Pa (0.3 in. of water). We have performed our calculation for three different values of coefficient C in Eq. (2). A value of C = 0.038 liter/sec·m^2·Pa$^{0.65}$ (0.47 in.^{3}/sec·in.^{2}·psi$^{0.65}$) is used for tight systems. For a tight system, the air flow rate is 0.63 liter/sec·m^2 (0.125 cfm/ft^2) at a pressure difference of 75 Pa (0.3 in. of water). The corresponding values of C used for average and loose systems are 0.093 and 0.183 in SI units (1.14 and 2.25 in units of inches, pounds and seconds).

Let us consider an inflated membrane either circular or long rectangular in plan. For a membrane subjected to internal pressure, the deflection w is a quadratic function of distance. The maximum deflection of the membrane may be represented in terms of the material properties of the membrane as follows [3]:

$$w_{max} = a\,\alpha\left[\frac{a}{Eh}\left(P_b - P_e\right)\right]^{\frac{1}{3}} \tag{3}$$

where α = a dimensionless coefficient that depends on the shape of the membrane in planform and can be shown to be equal to 0.91 $(1 - \nu^2)^{1/3}$ for a long rectangular and 0.72 $(1 - \nu)^{1/3}$ for a circular membrane with Poisson's ratio ν, a = radius for a circular membrane and one-half of the distance between supports in the short direction for a long rectangular membrane, E = modulus of elasticity of the membrane, and h = membrane thickness.

The volume under the inflated membrane may be estimated from the following equation [3]:

$$V_b = V_{bo} + \gamma\, A\, w_{max}$$

$$= V_{bo} + \gamma\, A\, \alpha\, a\left[\frac{a}{Eh}\left(P_b(t) - P_e(t)\right)\right]^{\frac{1}{3}} \tag{4}$$

where V_{bo} = the volume of trapped air between the membrane and roof deck when w_{max} = 0, A = the projected area of inflated membrane, and γ = a dimensionless coefficient equal to 0.67 for a long rectangular membrane and 0.25 for a circular membrane.

The three equations (1), (2), and (4) can be solved in terms of the unknowns $P_b(t)$, $V_b(t)$, and Q. Specifically, these equations can be simplified to give a single nonlinear differential equation in terms of $P_b(t)$, as follows:

$$\frac{d}{dt}\{P_b(t)[V_{bo}+ K(P_b(t) - P_e(t))^{\frac{1}{3}}]\} = CA[P_i - P_b(t)]^{0.65} \, Sgn[P_i - P_b(t)] \qquad (5)$$

where $K = \gamma \, A \, \alpha \, a \, (a/Eh)^{1/3}$ and Sgn (x) = 1 for x ≥ 0 and -1 for x < 0.

When the external wind pressure, $P_e(t)$, is a known function of time and when values of P_i, V_{bo}, C and A are known, the above equation can be solved numerically for $P_b(t)$; the pressure difference across the membrane thickness, $P_b(t) - P_e(t)$, is then calculated. When only the power spectrum of wind pressure on a roof, rather than its functional value with time, is known, the solutions of Eq. (5) to sinusoidal external wind pressure functions with different frequencies are used to approximate the rms of pressure fluctuations across the membrane thickness, $P_b(t) - P_e(t)$.

The tension in the membrane for both long rectangular and circular membranes is expressed in terms of pressure difference across the membrane thickness by

$$T = \frac{\beta \, a^2}{2 \, w_{max}} \left(P_b(t) - P_e(t) \right) \qquad (6)$$

and the vertical reaction at the boundary of the membrane measured in units of force per unit length of the perimeter of the inflated membrane by

$$R = \beta \left(P_b(t) - P_e(t) \right) a \qquad (7)$$

where $\beta = 1$ for a long rectangular membrane and 0.5 for a circular membrane [3].

To investigate the accuracy and validity of Eqs. (3), (6) and (7) governing the behavior of a membrane subjected to pressure, a finite element analysis of a 3.7 m by 7.3 m (12 ft by 24 ft) membrane attached with bars, 2.1 m (7 ft) apart, running parallel to the short direction is performed. This system, which had passed the underwriter's test for 4.3 kPa (90 psf) allowable wind pressure, was actually tested to failure; failure occurred at a pressure of 2.5 KPa (52 psf). In the finite element analysis performed, the membrane is assumed to be linearly elastic with E = 95.8 MPa (13,900 psi), h = 1.2 mm (0.048 in.) and ν = 0.4. The analysis accounts for the geometric nonlinearity of the deformed membrane. The analysis is performed on a quarter of the structure as shown in Fig. 2. The results of the finite element analysis are compared with the results obtained using Eqs. (3), (6), and (7) in Table 1.

The results show that the approximate equations presented here for the deflection of the membrane, the magnitude of tension in the membrane, and the vertical reaction are accurate.

TABLE 1 – Comparison of the results of finite element analysis of a pressurized 3.7 m by 7.3 m (12 ft by 24 ft) rectangular membrane with the simplified expressions, Eqs. (3), (6), and (7).

	Finite Element	Eqs. (3), (6), and (7)
Membrane Tension (N/mm)		
Center of Inflated Membrane	5.0	5.3
Maximum at Bar Attachment	5.9	5.3
Average at Mid-Width	5.5	5.3
Maximum Deflection (mm)	259	267
Maximum Vertical Reaction (N/mm)	5.5	5.3

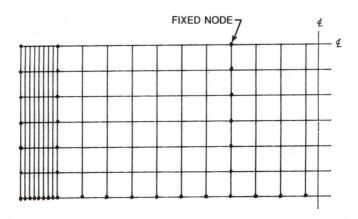

FIG. 2 – Finite-element model used for analysis of a pressurized membrane

WIND PRESSURE ON FLAT ROOFS

Wind pressure on a roof consists of a static component and a fluctuating component. The measurements of the static wind pressure on a rectangular-plan flat-roof building have been performed in wind-tunnel and reported as mean-hourly pressures. The fluctuating component of wind pressure has been often reported in the form of rms pressure or peak pressure, e.g., Stathopoulos [6], and Stathopoulos, Surry, and Davenport [7], Kind [8 and 9], and only recently in the form of power-spectra (see Wacker et al. [10]). The measurements of power spectra were conducted for a point at the corner and a second point near the edge of a medium-rise building subjected to a wind that makes a 45 degree angle with the facades. The power spectra for these two points are quite similar. The power spectrum of wind pressure at the corner point, S(f), expressed in terms of the frequency and the rms of the wind pressure fluctuations, f and σ, building width, W, and mean-hourly wind speed at roof height, $\bar{V}_h$, is shown in Fig. 3.

Based on the results of wind-tunnel tests performed by Stathopoulos et al. [5], the following conclusions can be made:

- The peak suction on flat roofs occurs for the 45 degree oblique wind near the upstream corner of the building.

- The magnitude of peak suction coefficient, measured relative to $q = (1/2) \rho \bar{V}_h^2$ the dynamic pressure corresponding to the mean-hourly wind velocity at the roof height, $\bar{V}_h$, depends on the height of the building and the type of terrain (smooth or built-up).

- The region of the high suction is larger for the higher-rise buildings.

- In the regions of highest suction, the ratio of the static-to-peak wind pressure ranges from 1/4 to 1/2 for a medium-rise building (comparable to 69 m (225 ft) height full scale on a 76 x 76 m (250 x 250 ft) building footprint) and 1/7 to 1/3 for the low-rise building (comparable to 15 m (50 ft) height full scale on a 76 x 76 m (250 x 250 ft) building footprint). The larger ratios are for a smooth terrain. (Significantly higher wind pressures over highly localized areas have been reported by Kind [8 and 9]; his results show that the ratio of the static-to-peak wind pressure for low-rise building is also close to 1/2.)

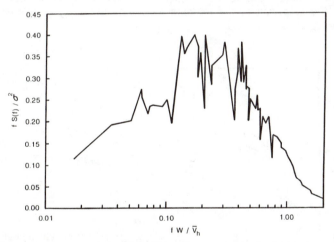

FIG. 3 – Power spectrum of the fluctuating component of wind pressure at the corner of a square-plan, flat-roof, medium-rise building measured in a wind-tunnel by Wacker, et al. [10].

RESPONSE TO STATIC WIND COMPONENT

Let us consider the response to the static wind suction, P_{eo}, for a membrane that allows air infiltration into the space underneath it. The membrane will inflate under the differential pressure $P_b - P_{eo}$. The initial volume of trapped air under the membrane is given by Eq. (4) with P_{eo} and P_i replacing $P_e(t)$ and $P_b(t)$, respectively. The membrane is subjected to a static tension that is calculated from Eq. (6), and the perimeter of the inflated membrane is subjected to a static component of vertical reaction, R, given by Eq. (7) with P_{eo} and P_i replacing P_e and P_b, respectively.

RESPONSE TO FLUCTUATING COMPONENT OF WIND PRESSURE

When a membrane, inflated by the static component of wind pressure, is subjected to the fluctuating component of external wind pressure, the pressure underneath the inflated membrane will change with the external wind pressure and, therefore, the pressure differential across the membrane thickness will be less than the fluctuating component of wind pressure. The ratio of the pressure differential across the membrane thickness to the fluctuating wind component of pressure is called the transient response factor. Zarghamee [3] calculated this response factor for assumed sinusoidal variations of external wind pressure. In this paper, this dynamic response factor is calculated for a wind-tunnel-simulated wind; the power spectra of wind pressure at the corners and the edges of a square-plan, flat-roof building, measured by Wacker et al. [10] in the wind tunnel of the University of Karlsruhe, are used to compute the transient response factor.

Calculations are performed for different values of $W/\bar{V}_h$, where W is the horizontal dimension of the side of the square building and $\bar{V}_h$ is the mean-hourly wind speed at the roof height. The results are shown in Fig. 4 for a point near the corner of a medium-rise, square plan, flat roof building in a smooth terrain. The results show that the dynamic response factor depends on the size of the building and the mean-hourly wind speed at the roof height. The larger the building width and the lower the mean-hourly wind speed at the roof height, the larger is the response of the membrane to the fluctuating component of wind pressure.

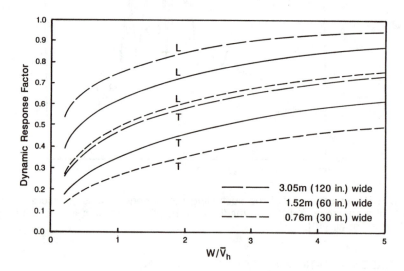

FIG. 4 - Transient response factor, i.e. the fraction of the fluctuating component of wind pressure across a 0.76 m (30-in.), a 1.52 m (60-in.) and a 3.05 m (120-in.) wide rectangular membrane with h = 1 mm (0.04 in.) and E = 172 MPa (25,000 psi), subjected to the wind-tunnel-measured power spectrum of Fig. 3 and with different air infiltration characteristics (i.e., T for tight and L for loose).

In addition, the results of Fig. 4 show that the larger the width of the membrane, the greater the response of the membrane to the fluctuating component of wind pressure. The reaction at the boundaries of the inflated membrane is proportional to the product of pressure difference across the membrane thickness and the size of the inflated membrane; therefore, the vertical reaction, R, grows more rapidly with the size of the inflated region than does the pressure difference across the membrane thickness. The results of Fig. 4 show also that the rate of air infiltration into the space under the membrane is very important in determining the response of the membrane to the fluctuating component of wind pressure.

DISCUSSION

Single-ply roofing systems with membranes that respond to wind suction by inflating must be designed for the static plus a part of the fluctuating component of wind pressure. The static component of wind pressure is usually measured as mean-hourly pressures in wind-tunnel tests performed specifically for a building. Mean-hourly pressures are also reported in the literature for buildings with simple geometries, usually with rectangular plan and flat roof with or without parapets. The fluctuating component of wind pressure can be obtained by subtracting the static component from the peak wind pressures measured in wind tunnel tests or from the design peak wind pressures calculated based on the existing standards. The fraction of the fluctuating component of wind pressure that should be considered in design depends upon the following factors:

1.   Rate of air infiltration into the space underneath the membrane

2.   Size of the building

3.   Maximum size of the inflated region or membrane width.

If air infiltration into the space underneath the membrane is unimpeded, the full fluctuating component of wind pressure should be used for design. Such a case can occur, for example, when a soft or loose membrane, discretely attached with washer plates to the roofing deck, is subjected to high wind. The membrane, inflated under the static wind pressure, may provide unimpeded air flow to the localized areas of the membrane where wind pressure is maximum.

Air infiltration can also occur from the perimeter of the membrane. Fig. 5 shows a missing end closure in the edge details of a curtain wall at the corner of a building. The positive pressure on the wall increased the rate of air infiltration into the space underneath the membrane and caused the failure of the roofing system. Note that in such a case, the effective wind pressure on the membrane may in fact exceed the peak wind pressure on the roof.

The size of the inflation area can be controlled by attachment details. The use of a bar attachment system arranged to (1) limit the area of inflation, and (2) prevent unimpeded air flow to the space underneath the inflated membrane, can dramatically reduce the fraction of the fluctuating component of wind pressure that should be accounted for in design.

FIG. 5 – Missing end closure causes increased rate of air infiltration.

If the attachment system and edge details do in fact impede air flow to the space underneath the membrane, then the fraction of the fluctuating component of wind pressure that should be used for design of the membrane and its attachment system is tentatively proposed Table 2. For this purpose, let us define tight, average, and loose systems as follows.

High Air Infiltration Rate: Plank wood decks, decks made of shredded wood, precast concrete panels, metal decks with large cracks, perforation, missing end closures, and openings to the interior of the building, and any roof deck with expansion or perimeter joints that permit easy flow of air into the space under the membrane.

Average Air Infiltration Rate: Wood and metal decks that have been specifically designed and built to reduce the expected normal air infiltration to the space between the membrane and roof deck. To this end, perforations and openings to the interior of the building are eliminated, end closures are in place and expansion and perimeter joints are sealed.

Low Air Infiltration Rate: Concrete decks with sealed expansion joints and with membrane perimeter details specifically designed to eliminated air infiltration.

TABLE 2 – Percent of dynamic wind pressure to be used for design

| | Air Infiltration Class | | |
Building Width	Loose	Average	Tight
≤ 15 m (50 ft)	60	50	40
30 m (100 ft)	70	60	50
≥ 60 m (200 ft)	80	70	60

These numbers are based on the power spectra measured by Wacker, et al. [10] and the calculated transient response factors shown in Fig. 4. In this framework, they provide conservative estimates for the fluctuating component of wind pressure that is applied to membranes with 1.5 to 1.8 m (5 to 6 ft) width. When the width of the membrane is about 0.6 to 0.9 m (2 to 3 ft), the above numbers may be decreased by 15 percentage points. (Note that the mechanical attachment controls the size of the inflated area.)

Infiltration also affects moisture migration to the space underneath the membrane and condensation; this issue should be considered in the design although it is outside of the scope of this paper.

CONCLUSIONS

The response of single-ply roofing systems to the fluctuating component of wind pressure on the roofs of rectangular-plan, flat-roof buildings is evaluated based on the reported power spectra of wind pressure measured in a wind-tunnel test. The results show that single-ply roofing systems should be designed for the static component plus a fraction of the fluctuating component of wind pressure. The magnitude of the design wind pressure depends on the rate of infiltration into the space underneath the membrane, the size of the building, and the size of the region of the membrane that inflates in wind. Specific recommendations are tentatively proposed for design wind pressures for single-ply roofing systems. Of course, more work is required, specifically on quantifying the rate of air infiltration to the space underneath the membrane and on measuring wind pressure spectra on roofs to verify and support such recommendations.

REFERENCES

[1]   Kind, R.J., and Wardlaw, R.L., "Wind Tunnel Tests on Loose-Laid Roofing Systems for Flat Roofs," Second International Symposium on Roofing Technology, 1985, pp. 230-235.

[2]   Kind, R.J., and Wardlaw, R.L., "The Development of a Procedure for the Design of Rooftops Against Gravel Blow-off and Scour in High Winds," NRCA/HBS Proceeding of the Symposium on Roofing Technology, Paper Number 16, 1977, pp. 112-123.

[3]   Zarghamee, M.S., "Wind Effects on Single-Ply Roofing Systems," Journal of Structural Engineering, ASCE, Vol. 116, No. 1, January 1990, pp. 177-187.

[4]   ASHRAE Handbook – 1985 Fundamentals, American Society for Heating, Refrigerating and Air Conditioning Engineers, Atlanta, Ga., p. 22-11.

[5]   Tamura, G.T., and Shaw, C.Y., "Studies on Exterior Wall Air Tightness and Air Infiltration of Tall Buildings," ASHRAW Transactions, Vol. 82, Pt. 1, 1976, pp. 122-134.

[6]   Stathopoulos, T., "Wind Pressure Functions for Flat Roofs," Journal of Engineering Mechanics, ASCE, 107, EM5, 1981, pp. 889-905.

[7]   Stathopoulos, T., Surry, D., and Davenport, A.G., "Effective Wind Loads on Flat Roofs," Journal of the Structural Division, ASCE, Vol. 107, ST2, February, 1981, pp. 281-298.

[8]   Kind, R.J., "Worst Suctions Near Edges of Flat Rooftops on Low-Rise Buildings," Journal of Wind Engineering and Industrial Aerodynamics, Vol. 25, No. 1, 1986, pp. 31-47.

[9]   Kind, R.J., "Worst Suctions Near Edges of Flat Rooftops with Parapets," Journal of Wind Engineering and Industrial Aerodynamics, Vol. 31, No. 2-3, pp. 251-264.

[10] Wacker, J., Friedrich, R., Plate, E.J., and Bergdolt, U., "Fluctuating Wind Load on Cladding Elements and Roof Pavers," Proceedings of the 8th Colloquium on Industrial Aerodynamics, Aachen, September 04-07, 1989, pp. 225-237.

MSZ18-89.ct

Jorge E. Pardo

WIND TEST METHODOLOGY FOR LOOSE-LAID ROOF ELEMENTS AND COMPONENTS

REFERENCE: Pardo, J. E., "Wind Test Methodology for Loose-Laid Roof Elements and Components", Roofing Research and Standards Development: 2nd. Volume, ASTM STP 1088, Thomas J. Wallace and Walter J. Rossiter, Eds. American Society for Testing and Materials, Philadelphia, 1990.

ABSTRACT: This report concerns an investigation into the wind performance limits of loose-laid roof elements, as exemplified by membranes, pavers and other components of ballasted single-ply roofing systems. On the basis of comparisons between small scale model tests in Atmospheric Boundary Layer (ABL) wind tunnels and full scale prototype tests in aeronautical wind tunnels, a test method is developed and a semi-empirical analysis procedure is presented to account for the effects of air turbulence upon the limit windspeeds observed.

KEYWORDS: wind uplift, ballasted roofs, low slope roofs, ballasting pavers, dynamic wind tests, microzone effects, wind turbulence

INTRODUCTION

The determination of the wind speed limits at which loose-laid roofing system elements initiate displacement has important economic and safety implications, since wind damage to building roofs and adjacent structures has been documented as being preceded by such displacement. (FM-Loss Prevention Data 1-29, 1984).

The first generations of ballasted single-plies utilized stone aggregate, preferably round and smooth, as gravity securement of the membrane to the roof substrate.

Aside from its economy in areas where abundant, aggregate ballasting has demonstrated a propensity for becoming an airborne missile when under extreme wind conditions, while under less severe conditions, it may be scoured by relatively low velocity winds. Under extreme circumstances, aggregate ballast scouring and pile-up may lead to deck overloading in areas of accumulation, or at best, necessitates periodic stone redistribution over bare spots of membrane.

Mr. Jorge Pardo is a member of the American Institute of Architects, the Wind Engineering Research Council, the Single Ply Roofing Institute Technical Committee, and Director of the Innovative Design Research Division of NCMA, P. O. Box 781, Herndon, VA 22070 - U.S.A.

For the above reasons, a number of roof paver systems have been developed for the ballasting of single-ply roofs, in which desirable fea-
tures may be incorporated insofar as freeze-thaw resistance, drainage,
interlocking, and color and texture variations.

Background

Building macro-structures (entire buildings and building clusters)
are currently tested using reduced-scale models in meteorological wind
tunnels, where the natural boundary layer characteristics of the atmospheric wind can be simulated in regards to variation of speed and
turbulence with height. (Cermak, 1977). This method is essential for
the accurate study of building macro-structure behaviour under high
winds, and is hereinafter referred to as "Macro-zone" wind testing.
(Figure 1).

The wind testing of building roof Micro-structures on the other
hand, (subcomponents and elements), such as the loose-laid assemblies
scrutinized in this study, is more advantageous when performed on full
scale specimens. This approach eliminates compromises in model characteristics, like thickness/density ratios, membrane elasticity, and air
permeability of the deck, to name a few. Such compromises are usually
required to feasibly downscale microstructure elements for testing in
atmospheric boundary layer (ABL) type tunnels. (Kind, 1979).

Low slope roofing assemblies have traditionally been tested for
wind uplift through the application of static pressure differentials
across full scale sections of the prototype specimen. (FM-4470/UL-580).
The configuration and nature of loose-laid roofing systems preclude
this type of static-load testing for their evaluation, as the displacement and uplift of loose-laid assemblies are the resultant of dynamic
airflow conditions on the roof surface, to which underdeck infiltration
contributes as discussed later.

In view of the preceding, the concept of "Micro-zone" wind testing
has been advanced (Pardo, 1986) in light of Kind & Wardlaw's conclusion that the only wind characteristics immediately relevant to the
wind-induced displacement of roof micro-structures are those encountered at roof-top level. (Kind, 1979). The micro-zone method developed
for this study, (Figure 2) allows testing of roof microstructures in
aeronautical wind tunnels under relatively smooth-flow conditions, and
independently of their location within a (simulated) boundary layer.

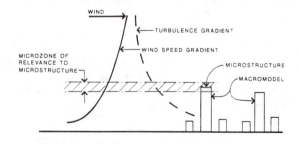

FIG. 1 --   **MACROZONE WIND TEST**
ABL WIND TUNNEL

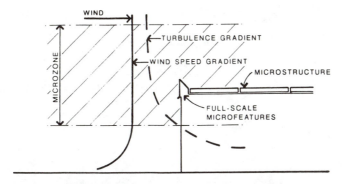

FIG. 2 -- **MICROZONE WIND TEST**
AERONAUTICAL WIND TUNNEL

Experiments: The tests described in this study were conducted at the Glenn L. Martin Wind Tunnel facility of the Aerospace Engineering Department of the University of Maryland. The test facility is an aeronautical, closed-loop wind tunnel with a 7.75x11.04 ft. (2.4x3.4 meter) test section at which point a maximum air speed of 230 mph (370 kph) is attainable. Accurate velocity/pressures can be maintained in the tunnel with the 2,000 hp synchronous motor driving a fixed-pitch blade fan.

Turbulence levels for this study were measured at the leading edge of the test table utilizing hot film anemometers, and flow speed measurements were taken via Pitot-static tubes at two levels near the point of incidence, with Gould type 590 Barocel differerential pressure transducers comparing total pressure, as detected in the settling chamber, to the static pressure at the pitot tube locations.

Static pressures from 148 taps on the test rig were fed to a mini-computer for digitization, via Scanivalves and pressure transducers, interconnected by means of .054" I.D. tubing for relatively flat amplitude response in the 10 to 20 HZ range.

The microzone wind test method allows the precise duplication of typical roof deck assemblies (Figure 5) in which metal deck, insulation, single-ply membrane and fascia elements can be put together as they are in actual buildings, thereby replicating the relative air permeability of this type of roof deck construction.

Metal decking utilized for the model roof construction complied with the specific requirements of Factory Mutual Engineering Corp. (insofar as physical characteristics and securement to the roof structure) as applicable to Class I type steel deck roofs.

Undermembrane rigid insulation 1-1/2" thick (3.8 cm) and 1/2" (1.27 cm) gypsum board fire-barrier (used on PMR tests only) were laid loose, and then covered by a 45 mil. EPDM impermeable membrane also laid loose, except for perimeter anchorage.

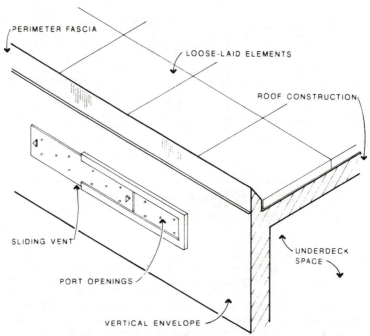

FIG. 3 -- **VARIABLE INFILTRATION PORTS**
FOR TEST TABLE UNDERDECK PRESSURIZATION

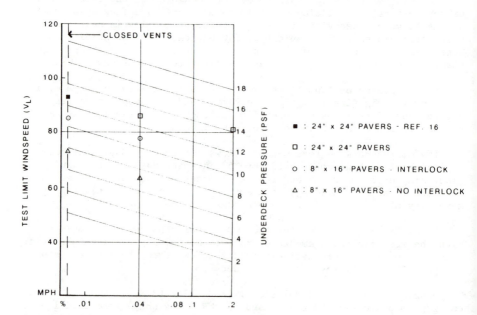

■ : 24" x 24" PAVERS - REF. 16

□ : 24" x 24" PAVERS

○ : 8" x 16" PAVERS - INTERLOCK

△ : 8" x 16" PAVERS - NO INTERLOCK

GRAPH 1 -- **EFFECT OF ENVELOPE INFILTRATION**
BALLAST PAVERS - WIND UPLIFT

The ballast pavers tested included type "R" of 8"x16"x2.5" (20x40 x6.3 cm) in plan exerting an average ballast pressure of 12.7 lb/s.f. (608 N/m^2). The larger pavers type "P", used as reference, were approximately 24"x24"x2" (60x60x5cm) and exerted a ballasting pressure of 23.75 lb/s.f. (1136 N/m^2).

Underdeck Pressurization: As discussed in the literature, the effects of underdeck air pressure resulting from building envelope openings and/or infiltration, can be a major contributing factor to the overall uplift pressure exerted upon roof coverings by high winds, and consequently, on the limit airspeed which such coverings can stand before displacement. (FM-Loss Prevention Data 1-7, 1983). For this reason, the correlation between different envelope infiltration areas and limit windspeed was investigated in this study by means of a set of variable vent ports located on two opposite sides of the test table. By operation of these ports, openings corresponding to specific percentages of the roof deck area could be vented to the incoming flow, or conversely closed totally. (Figure 3).

In this manner, the type of roofdeck construction could be simulated, as it is known that monolithic cementitious roof decks tend to isolate well the roof coverings from the pressures which may result inside the building, while metal roof deck construction is rather permeable to internal building pressurization, bringing about correspondingly deleterious effects upon the wind uplift performance of the roof coverings. (Graph 1).

While the research program reported here did not include infiltration measurements of the roof deck itself, the air permeability of a typically constructed metal roof deck assembly (Figure 5) was quite apparent by the direct impact evident upon test results when varying the opening of the envelope infiltration ports, which illustrated additionally, that the transparency of metal deck construction, to air passage, can be reduced through the use of air barriers such as gypsum board.

Interchangeable Fascia: It has been noted that reduced scale models (1:10 to 1:15) utilized for wind uplift research of microstructures grossly depict fascia/membrane interphase conditions, and as a rule, such highly abbreviated parapet configurations unrealistically protect membrane edges from air infiltration. Additionally, vortex formation and other air disturbances resulting from wind impact upon the roof perimeter are highly dependent upon the geometry of the fascia micro-structure, as evidenced by the results of this study; a fact usually dismissed in macrozone-type tests.

Four different fascia configurations were tested: a- A 6" (15cm) straight fascia; b-6" (15 cm) straight fascia with retainer strip; c- A 3" (7.6cm) straight fascia, and d-A 6" (15cm) experimental type de-eloped specifically for these tests. (Figure 4).

Removable perimeter frames were fitted with fascia assemblies on three (3) sides only, while the fourth side was left flush with the surface of the loose-laid elements under test, and fitted with especially shaped metal trim, designed to retain the loose-laid elements laterally in the case of non-interlocked arrays, and in the case of interlocked arrays, to simulate the vertical restraint corresponding to the interlock provided by a continuing (virtual) series of elements, as would occur in a much larger roof surface.

Gauge Deck: Previous research has shown that the pressure-equalization phenomenon-which occurs across top and bottom surfaces of loose-laid roof elements-is related to joint configuration and joint length density per unit area.

Pressure equalization is an important parameter in the uplift behaviour of roof pavers and boards, in that it substantially affects the net forces "sensed" by loose-laid elements, for the negative pressures acting on their top surfaces tend to be cancelled by the negative pressures acting underneath, as transmitted quasi-instantaneously through the joint grid.

In order to observe the pressure equalization effect on loose-laid elements of different aspect ratios and with different joint densities per unit area (i.e. different sizes), an elevator-type gauge deck assembly was designed and constructed to operate as illustrated in Figures 6 & 7.   The movable gauge deck consisted of a wood platform fitted with a nodal grid of pressure taps spaced at 8" (20cm) in either direction, feeding scanivalves as discussed earlier.

Gauge deck operation permitted measurement of pressures at precise roof elevations in respect to the fascia height, such as corresponding to the paver top surfaces, and at the location of the roofing membrane, with actual paver arrays laid upon it, in order to measure under-element pressures.

Precise elevation of pressure sensing surfaces in respect to fascia height is important because it has been demonstrated (Kind, 1979), (Bienkiewicz & Meroney, 1985) that fascia height has a direct bearing upon the magnitude of the negative pressures acting on the roof (components) surface.

The novel concept of a movable gauge deck for roof pressure measurements is considered an advance over the methodologies reported in the literature (Kind, 1979), in which a fixed gauging surface is utilized to scan pressure taps with, and without roof elements upon it, because such method obviates measurements at two different fascia heights (i.e. top and bottom of pavers).

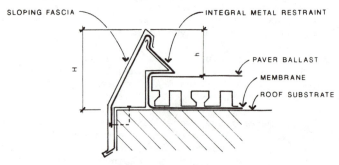

FIG. 4 --   **PERIMETER C**
RESTRAINED

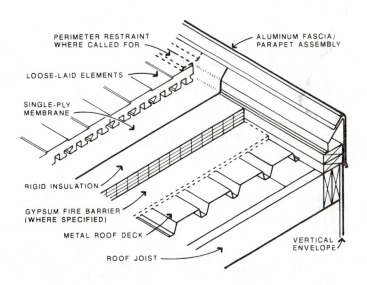

PERIMETER RESTRAINT
WHERE CALLED FOR

ALUMINUM FASCIA/
PARAPET ASSEMBLY

LOOSE-LAID ELEMENTS

SINGLE-PLY
MEMBRANE

RIGID INSULATION

GYPSUM FIRE BARRIER
(WHERE SPECIFIED)

METAL ROOF DECK

ROOF JOIST

VERTICAL
ENVELOPE

FIG. 5 -- **MODEL ROOF CONSTRUCTION**

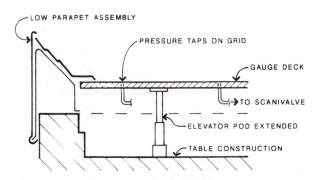

LOW PARAPET ASSEMBLY

PRESSURE TAPS ON GRID

GAUGE DECK

TO SCANIVALVE

ELEVATOR POD EXTENDED

TABLE CONSTRUCTION

FIG. 6 -- **GAUGE DECK UP**
FOR TOP SURFACE MEASUREMENTS

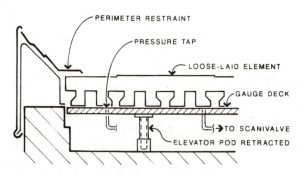

FIG. 7 -- **GAUGE DECK DOWN**
FOR UNDER-ELEMENT MEASUREMENTS

Experimental Procedure: The wind studies referenced here indicate
that the most sensitive areas of the roof, to wind-induced damage, are
near the building corners, particularly when the wind flow originates
at 45° from the windward corner thereby bisecting it. (Kind, 1976).

The tests described in this report explored the critical 45° wind
direction identified, and made use of a rotating test table to vary
prototype symmetry parameters.

Each experiment was conducted by gradually increasing the tunnel
wind-speed. Upon attainment of limit conditions, (as defined under
"Limit Windspeeds" below) air velocity was maintained constant for a
minimum of three (3) minutes in order to observe periodic motions of
the system under scrutiny, which might lead to instability or fatigue
failure.

All experiments were videotaped, while visual monitoring noted
events of particular interest; tunnel airspeeds were recorded at limit
cases, and at other relevant times as described further in the full
study. (Pardo, 1988)

Limit Windspeeds: The critical windspeed identified for the purposes of this report as (V_L), or limit windspeed, is defined here as
the condition at which the measured mean tunnel windspeed is sufficient to effect uplift at any one point, of any single roof element, by
a maximum of 2" (5 cm), for a minimum period of three (3) minutes.

Vertical displacement of less than 2" is allowed, as it results in
fluttering, ballooning and similar motions of the assembly, but without
developing the instabilities which lead to dislodgement.

The experience gathered from the experiments described by Pardo
(1986) confirms that, for the type and size of loose laid elements under consideration, which typify all paver/board designs currently in
use in the U.S., this amount of vertical displacement accurately fore-
tells proximity of failure, which in turn is defined as element dislodgement from the assembly.

EFFECTS OF TURBULENCE

One of the findings of the research by Bienkiewicz and Meroney (1985) is that a higher degree of longitudinal turbulence in the roof-top air flow results in a lowering of the mean windspeeds required to effect uplift of roof elements.

A semi-empirical relationship derived from such tests has been developed by the writer as a preliminary step to account for the effects of air turbulence upon the limit windspeeds governing uplift of loose-laid roof assemblies.

Unrestrained, Loose-laid Roof Elements: For unrestrained i.e. non-interlocked roof assemblies, the relationship between turbulence intensity and limit windspeed (Graph 2) may be expressed as follows:

$$V_{LX} = V_{LZ} \{1 + [E(t)_x - E(t)_z] \}^{1/2} \tag{1}$$

where,

V_{LX} = Limit windspeed at turbulence sought
V_{LZ} = Limit windspeed at reference turbulence
$E(t)$= Turbulence effect factor, defined as,

$$E(t) = (1/I_u)^{1/2} \tag{2}$$

where,

I_u = longitudinal turbulence intensity (%)

and subscripts,

Z   = reference condition
X   = condition sought

Aeronautical Wind Tunnel Tests: Since the dynamic effects of turbulence on roof element limit windspeeed (re: above) (V_L) below approximately (I_u) 5% can be considered as negligible, the conversion of test results for unrestrained assemblies conducted in aeronautical tunnels to probable performance levels at higher turbulences, makes use of a simplified version of Equation (1) in the form of:

$$V_{LX} = V_{LZ} \{ 1+ [E(t)_x - .5] \}^{1/2} \tag{3}$$

Interdependent, Loose-laid Assemblies: Graph 2 illustrates that in the case of perimeter-restrained, interlocked assemblies, the relationship between the uplift wind speed limits and the rooftop airflow turbulence, develops a steeper slope than for unrestrained assemblies, as predicted by the following expression:

$$V_{LX} = V_{LZ} \{(I_{uZ}/I_{ux})^{1/2} + [E(t)_x - E(t)_z] \}^{1/2} \tag{4}$$

On the basis of the rather limited data available, the above equation appears to be applicable for the conversion of limit windspeeds of restrained assemblies within a turbulence range of 7 to 20% (I_u). Method of application to aeronautical wind tunnel test results is given in the full study. (Pardo, 1988).

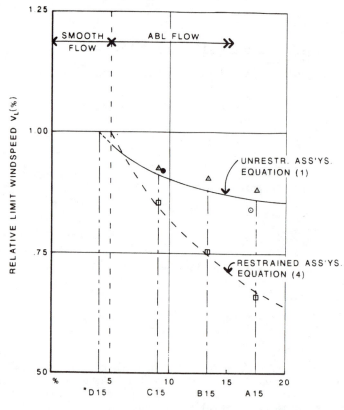

GRAPH 2 —   LONGITUDINAL TURBULENCE (I_u)

EFFECT OF TURBULENCE ON V_L
FOR LOOSE-LAID ROOF ELEMENTS

*Approx. turbulence at 15 ft. ht. for Exposure Type given by ANSI A-58.1

O = REFERENCE 8 (Not Interconnected)
● = REFERENCE 8 (Not Interconnected)
△ = REFERENCE 2 (Not Interconnected)
□ = REFERENCE 2 (Interconnected)

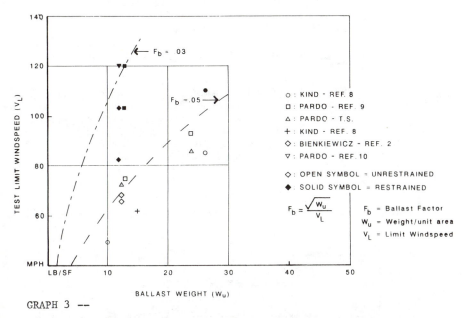

GRAPH 3 --

RESTRAINED vs. UNRESTRAINED ARRAYS
BALLAST PAVER WIND TEST RESULTS   6" PARAPET HEIGHT

RESULTS AND DISCUSSION

The adjusted uplift windspeed values (windspeed at which unstable roof element uplift is likely to occur) presented in Tables 1 and 2 allow direct comparison of tests conducted under airflows with different energetic content (turbulence), and reveal definite performance trends likely to correspond with probable behaviour of roof covering elements in actual building locations.

Generally, it can be noted that the adjusted uplift windspeed values derived from tests in relatively smooth airflows under the microzone approach, are more conservative than those obtained from macrozone tests. As discussed before, this owes to the fact that more realistic, and severe, conditions can be simulated with full scale element models, specifically in regards to deck permeability, the effects of perimeter (flashing) configuration, and undermembrane leakage.

In the case of unrestrained assemblies, the analysis of test results indicates that loose-laid pavers will meet stability limits as defined above (uplift) at windspeeds, which increase as their weight increases (Graph 3), and decrease in a rather straight-forward mode as the energy content of the fluctuating component (longitudinal turbulence) of the windstream increases (Graph 2). The introduction of under-deck pressurization results in a proportional decrease of the limit windspeeds attainable. (Graph 1).

Lowering the effective parapet height, as reported by Pardo (1988), or modifying the fascia geometry, also shows definite performance effects in terms of lessened resistance to wind uplift as evidenced by the pressure measurements of the experimental sloping fascia, which indicate generation of higher negative pressures than for the conventional straight perimeter.

As to restrained (interlocked) assemblies, it is apparent that a substantial increase in uplift performance can be expected vis-a-vis unrestrained installations, accompanied by somewhat lessened sensitivity to changes in underdeck pressurization (Graph 1), as the homogeneity of the interdependent assembly diminishes the tendency to balloon exhibited by loose-laid membranes under high winds.

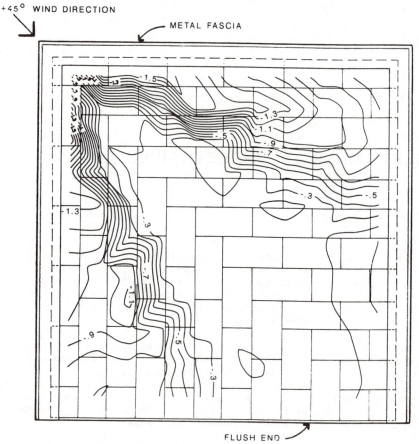

GRAPH 4 --

PRESSURE DISTRIBUTION   (Cp)*

TOP SURFACE OF PAVERS

*Nondimensional

PERIMETER : TYPE "A" FASCIA

WINDSPEED : 80 MPH

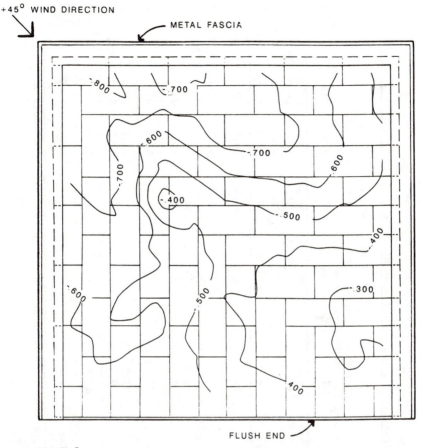

+45° WIND DIRECTION

METAL FASCIA

FLUSH END

GRAPH 5 --

PRESSURE DISTRIBUTION   (Cp)*

BOTTOM SURFACE OF PAVERS

*Nondimensional

PERIMETER : TYPE "A" FASCIA

WINDSPEED : 80 MPH

CONCLUSIONS

The nascent condition of research, regarding the effects of atmospheric turbulence upon the inertial characteristics of microstructures
in roof systems, is reflected in a modest body of data, which although
indicative of major relationships, needs to be expanded through further
investigation. The methods presented here to test microstructures in
aeronautical wind tunnels, and to convert the results to probable limits at atmospheric turbulence levels, although preliminary, should
prove valuable in the evaluation of different systems and their applicability within the wind exposures identified by the standards.

TABLE 1

Conversion of Wind Uplift Test Results to Standard Exposure*
For Unrestrained, Loose Laid Roof Elements

Reference (Test No.)	Micro-structure type	Roof Top test turbulence % I_u	Uplift** Windspeed Reported (mph)	[a]Adjusted Uplift Windspeed (mph)		
				A15	B15	C15
8 (22.1)	26 psf pavers	17.0	82	81.9	83.3	85.6
8 (22.3)	"	17.0	77.5	77.4	78.7	80.9
8 (83.18)	"	9.5	85	81.3	82.8	85.3
9 (7)	23.75 "	0.3	93	79.9	81.8	84.8
8 (F-4)	15 psf pavers on insul. boards	17.0	60	59.9	60.9	62.6
8 (F-7)	15 psf pavers	17.0	62	61.9	63	64.7
9 (1)	12.7 psf pavers	0.3	75	64.5	66.0	68.4
10 (24)	12.2 " " on insul. boards	1.16	80	68.8	70.4	72.9
T.S.[b] (7a)	12.2 psf pavers	1.16	73	62.7	64.2	66.5
2 (N.A.)	12 psf pavers	17.5	66	66.0	67.1	69.0
2 (N.A.)	" "	9.1	69.3	66.0	67.3	69.3
8 (22.4)	10 psf pavers	17.0	50	49.9	50.8	52.2
8 (x-1)	4.5 psf insul. boards	17.0	46	45.9	46.7	48.0

a   Via Equations (1) and (3) - For Conditions at Exposure type given, 15 ft. Ht.(Ref. 1)
b   T.S. = This Study.
*   Exposure as defined by ANSI A-58.1
** Stability limit

TABLE 2

Conversion of Wind Uplift Test Results to Standard Exposure*
For Restrained, Loose Laid Roof Elements

Reference (Test No.)	Micro-structure type	Roof Top test turbulence % I_u	Uplift** Windspeed Reported (mph)	[a]Adjusted Uplift Windspeed		
				A15	B15	C15
8 (83.25)	26 psf pavers strapped	9.5	110	88.8	98.1	111.6
T.S.[b](4b)	23.75 " pavers-perim.restr.	1.2	86[c]	73.9	75.7	78.4
10 (27)	12.2 psf pavers on insul. bds. (full coverage)	1.21	120	79.0	88.6	102.4
10 (31)	12.2 psf pavers on insul. bds. (partial coverage)	1.20	110	72.4	81.2	93.9
9 (4)	12.7 psf pavers	0.3	120	79.0	88.6	102.4
9 (3)	"	0.3	103	67.8	76.1	87.9
T.S.(43)	12.2 "	1.20	86[c]	56.6	63.5	73.4
2 (N.A.)	12 "	17.5	82.4	82.4	89.6	100.2
2 (N.A.)	"	9.1	103.7	82.2	91.0	103.7
8 (83.34)	4.5 psf. Insul.Boards	9.5	85	68.6	75.8	86.2
8 (83.33)	"	9.5	85	68.6	75.8	86.2
8 (83.31)	"	9.5	70	56.5	62.4	70.1

[a] Two-step conversion for values derived from interlocked element tests in flows below.
  5% I_u Equations (3) & (4) - Exposure type per Reference 1, 15 ft. ht.
[b] T.S. = This Study.
[c] High underdeck infiltration condition
* Exposure as defined by ANSI A-58.1
** Stability limit

The results of this test program are consistent with the data found in the literature, particularly in terms of the windspeed ranges at which loose-laid roof elements initiate (unstable) displacement. Despite the similarity in uplift results vis-a-vis macrozone tests, the pressure coefficients reported here appear to be lower than those measured under macrozone test conditions. The apparent discrepancy may be explained by bearing in mind that the underdeck (positive) pressure resulting from a permeable microzone test rig contributed (as in actual buildings) to the overall pressure acting on the loose-laid roofing systems being evaluated, and that a substantial pressure equalization effect was apparent across the gauge deck, with make-up air provided through the test table envelope.

The general behaviour of loose-laid roof assemblies under high laminar winds described here indicates that test samples of reduced dimensions - in the order of 72 sf $(6.7m^2)$ - illustrating rectangular roof corners, are capable of generating meaningful results for the wind uplift performance evaluation of microstructures on low slope roofs.

ACKNOWLEDGEMENTS

The author is indebted to Drs. J. B. Barlow and A. Winkelman respectively of the G. L. Martin Windtunnel and the Aerospace Engineering Dept. of the University of Maryland, for their assistance in setting up and interpreting the tests, and to A. Kassaee and W. Sekscienski of the G.L.M. Wind Tunnel who executed the tests and collected the data.

REFERENCES

[1] American National Standards Institute Inc., American National Standard A 58.1, 1982.
[2] Bienkiewicz, B., Meroney, R. N., "Wind-Tunnel Study of Westile Ballast Paver", Fluid Mechanics and Wind Engineering Program, Fluid Dynamics and Diffusion Laboratory-Civil Engineering Dept. Colorado State University CSU Project 2-96460 Report CER85-86BB-RNM13, Nov. 1985.
[3] Cermak, J. E., "Wind Tunnel Testing of Structures," Journal of the Engineering Mechanics Division ASCE, Vol. 103 No. EM6, Proc. Paper 13445, December 1977, pp.1125-1140.
[4] Factory Mutual Engineering Corp., "Wind Forces on Buildings and other Structures" in Loss Prevention Data, 1-7, 1983.
[5] Factory Mutual Engineering Corp., "Insulated Steel Deck" in Loss Prevention Data, 1-28, 1983.
[6] Factory Mutual Engineering Corp., "Loose-laid Ballasted Roof Coverings" in Loss Prevention Data, 1-29, 1984.
[7] Kind, R. J., Wardlaw, R. L., Design of Rooftops Against Gravel Blow-off, National Research Council of Canada, Report NRS 15544, September 1976.
[8] Kind, R. J., Wardlaw, R. L.,"Model Studies of the Wind Resistance of two loose-laid Roof-insulation Systems", National Research Council of Canada, NAE Report LTR-LA-234, May 1979.
[9] Pardo, J.,"The Effect of Variable Inter-connection on the Uplift Capacity of Roof Pavers" N.C.M.A. Report RCP-86702, July 1986.
[10] Pardo, J.,"Wind Performance Limits of Loose-Laid Roof Insulation Board Systems" N.C.M.A. Report 88950-686, February, 1988.

Bituminous Roofing Systems

Carl G. Cash[1]

POROSITY OF GLASS FIBER FELTS USED IN BUILT-UP ROOFING

REFERENCE: Cash, C. G., **"Porosity of Glass Fiber Felts Used in Built-Up Roofing",**Roofing Research and Standards Development: 2nd Volume , ASTM STP 1088, Thomas J. Wallace and Walter J. Rossiter, Eds., American Society for Testing and Materials, Philadelphia, 1990.

ABSTRACT: This paper describes a test method for measuring the relative porosity of asphalt coated glass fiber felts. For this paper, porosity is defined as flow through felt.

I tested the effect of the relative porosity of the felt using a computer model to calculate the flow of coal-tar pitch through the felts in an ideal membrane, exposed to normal Boston, Mass. and Miami, Fla. temperatures.

These data show the relative porosity of the glass fiber felts is highly variable. The mass of the built-up roofing aggregate presses the glass fiber felts to the bottom of the membrane.

I recommend, based on my field observations, our laboratory tests and the calculations reported in this paper, that only steep asphalt, ASTM D 312, Types 3 or 4, be used between the plies of asphalt coated glass fiber felts in the construction of built-up roofing membranes. Coal-tar pitch and dead level asphalt should not be used with glass fiber felts until dependable test methods and standards for the relative porosity of the glass fiber felts are established.

KEY WORDS: built-up roofing, glass fiber felts, coal-tar pitch, porosity, cold flow.

Introduction

I have frequently heard rumors of glass fiber felts "floating" or "sinking" during normal rooftop exposure. My own field observations of glass fiber felts in coal-tar pitch glass fiber membranes sinking to the bottom of the membranes confirm at least the sinking rumors, and my field observations are supported by the similar settling of glass fiber felts in accurately prepared laboratory membranes placed under heat lamps.

[1]Principal, Simpson Gumpertz & Heger Inc., consulting engineers, Arlington, Mass. 02174

The pitch cold flowing through the asphalt coated felts, as the glass fiber felts are pressed down by the mass of the aggregate, picks up asphalt and I observed the greasy and "foamed" interply bitumen, consistent with contact incompatibility.

Use of coal-tar pitch on asphalt coated glass fiber felts may result in an unbalanced membrane (all the reinforcement on the bottom of the membrane), deteriorated interply and top bitumen coatings (because they are at least partly composed of the products of contact incompatibility), cracked top coating, porous interply bitumen, and water leakage into the building.

This interply bitumen migration and deterioration has not been a problem when using coal-tar saturated felts. The most obvious difference between glass fiber and organic felts is the higher, but unquantified porosity of the glass fiber felts.

The following outlines a test method developed in our laboratory to measure the relative porosity of felts used in built-up roofing membranes. Relative porosity is defined as the volume of water per second flowing through the test felt under the arbitrary test conditions. This physical measurement does not take into consideration any of the more complex questions of incompatibility; contact incompatibility is beyond the scope of this paper.

Procedure:

We bolted felt samples between gasketed pipe flanges. The upper flange has a pipe stub to form a reservoir. Tap water is maintained at a constant head during the test interval, and the water flowing through the felts is collected in a tared container. The test interval most frequently used was 60 seconds, but used some intervals of over 1.5 hours (5400 s).

The results of each test, expressed as mean grams per second for the testing period is converted to mL/s x m^2 (in.^{3}/s x ft^2) by multiplying grams per second by 103.34 (0.5871).

The dimensions of our experimental apparatus are:

- effective sample diameter: 111 mm (4.37 in.)
- effective sample area: 0.0097 m^2 (0.1042 ft^2)
- head: 148 mm (5.8 in.)
- tap water temperature: 19°C(66°F)

We measured the repeatability of the test at two test intervals, the "normal" variation in the relative porosity of an ASTM D 2178, Type VI felt from one manufacturer, and the preliminary relative porosity of ASTM D 2178, Types IV and VI felts from five felt manufacturers vs. the relative porosity of asphalt-organic felt, D 226, Type 1.

We measured the viscosity of coal-tar pitch ASTM D 450, Types 1 and 3 from four pitch manufacturers using a parallel plate viscometer ASTM D 4989, and calculated the best least squares fit representing the linear relationship between the log-log of the viscosity and the temperature. We selected the parallel plate method because it uses relatively low shear rates and the viscosity is measured at a temperature range of 25° to 60°C (77° to 140°F), consistent with normal roof top environmental temperatures.

I used these data in a computer program to calculate the theoretical flow of the coal-tar pitch through the glass fiber felts in an "ideal" built-up roofing membrane exposed horizontally to environmental temperatures consistent with normal temperatures in Boston, Mass. and Miami, Fla. In part, I used a computer program (previously reported)[2] to generate the"normal"temperatures for a grey surface for each hour of each day, and arbitrarily selected the time needed for the mass of the coal-tar pitch between the bottom ply and the substrate to reach zero as the end point for most of the computer calculations.

Porosity Test Repeatability And Variation Within a Sample

We measured the water flow through one sample of glass fiber felt 24 times to get some estimate of the repeatability of the test method. Twelve of the tests were for 60 seconds, and twelve of the tests were for 120 seconds, to test the effect of doubling the test interval. Table 1 shows the relative porosity data from this test series.

These data show a statistically significant variation, at a 95% confidence level, between the average porosity measured using a 60 second test interval and a 120 second test interval. This variation may not be important considering the larger standard deviation in the data shown in Table 2, where 20 different samples of the same grade of glass felt by one manufacturer were tested.

TABLE 1 - Repeatability of Relative Porosity Measurements

Testing Interval:	60 Seconds		120 Seconds	
	g/s	Porosity	g/s	Porosity
(Porosity is shown in	34.02	3516	30.14	3115
mL/s x m²)	34.38	3553	30.48	3150
	35.98	3718	25.38	2623
	34.97	3614	26.83	2773
	32.60	3369	28.78	2974
	35.65	3684	29.09	3006
	33.72	3485	28.61	2957
	37.40	3865	30.38	3139
	33.55	3467	30.93	3196
	31.60	3266	31.73	3279
	30.75	3178	30.72	3175
	30.43	3145	30.82	3185
Mean		3488		3048
Standard Deviation		220.0		192.0
Mean Variance		4030		3072
Effective Degrees of Freedom			24	
Students' t (0.975)			1.99	
Expected Variation			168	
Mean (60 s) - Mean (120 s)			441	
		(Difference is significant)		

[2] Cash, C. G., "Computer Modeling of Climates," Insulation Materials, Testing, and Applications, ASTM STP 1030, D. L. McElroy and J. F. Kimpflen, Eds., American Society of Testing and Materials, Philadelphia, 1989, pp 599-611.

TABLE 2 - Variation in Porosity of Glass Felts
One Grade, One Manufacturer

Porosity mL/s x m^2			
3221	1114	2470	2175
1529	2568	1418	827
2173	1868	1643	2122
2684	1405	3857	2770
2186	3665	2260	1653
Mean			2180
Standard Deviation			780

Porosity Data

Table 3 shows the mean water flow, the porosity of seven glass fiber felts and one sample of perforated asphalt-organic felt obtained from five manufacturers. Each datum is the mean of six individual tests.

TABLE 3 - Water flow through built-up roofing felts

Felt Source	ASTM Type	Water Flow Rate, mL/s x m^2 (in.^{3}/s x ft^2)				Statistical Coefficients (log(Flow) = b - m(plies))		
		1 Ply	2 Plies	3 Plies	4 Plies	R	(m)	(b)
E	D 226 I	0.44	0.36	0.12	0.20	-0.759	-0.15	0.229
		(0.0025)	(0.0020)	(0.007)	(0.0011)	-0.785	-0.153	2.47
F	D 2178 IV	243	14.0	6.76	-	-0.946	-0.779	3.01
		(1.38)	(0.0792)	(0.0381)	-			0.766
	D 2178 VI	6990	1960	1110	952	-0.934	-0.285	4.00
		(39.6)	(11.1)	(6.31)	(5.40)			1.75
G	D 2178 IV	2280	1220	477	244.0	-0.998	-0.344	3.72
		(12.9)	(6.89)	(2.71)	(1.27)			1.43
	D 2178 VI	7600	3060	2020	1440	-0.971	-0.235	4.04
		(43.1)	(17.4)	(11.5)	(8.14)			1.80
H	D 2178 IV	10800	6580	4560	2840	-0.998	-0.190	4.22
		(67.2)	(37.3)	(25.8)	(16.1)			2.01
I	D 2178 IV	13100	7130	3900	2590	-0.996	-0.237	4.34
		(74.2)	(40.4)	(22.1)	(14.7)			2.09
	D 2178 VI	2180	1160	837	308	-0.979	-0.269	3.63
		(12.4)	(6.57)	(4.75)	(1.75)			1.38

As a matter of general interest, we tested the relative porosity of one through four plies for most of the felts, but I used the porosity of only one ply of felt in my computer program to estimate the effect of the porosity on the bitumen flow within a built-up roofing membrane. The empirical relationship between the porosity and the number of felt plies appears to be logarithmic.

The mean porosity measured varies from a low of 0.44 mL/s x m^2 (0.0025 in.^{3}/s x ft^2) for the perforated asphalt-organic felt, to 243 mL/s x m^2 (1.38 in.^{3}/s x ft^2) for the type IV felt from source F, to the 13,100 mL/s x m^2 (74.2 in.^{3}/s x ft^2) for the type IV felt from source I. Within the seven glass fiber felts tested, the felt with the highest porosity is more than 50 times as porous as the glass felt with the lowest porosity.

We have not tested all of the glass felts on the market, and our tests are not necessarily representative of the products offered by these manufacturers. The extreme variation in the porosity of these randomly selected glass felts probably affects the performance of the built-up roofing membranes that utilize these felts. The balance of this paper explores the effect of the porosity on the stability of the felts within an "ideal" built-up roofing membrane.

TABLE 4 - Temperature - Viscosity Curve Data, Coal Tar Pitch

Source:	A	B	C	D	A	C	D
Type:	1	1	1	1	3	3	3

Temperature
$^{\circ}$C ($^{\circ}$F)   Parallel Plate Viscosity, Pa x s (cps)

	A	B	C	D	A	C	D
25	3.3E+08	4.2E+08	2.8E+08	2.7E+08	3.4E+08	7.4E+08	2.4E+09
(77)	(3.3E+10)	(4.2E+10)	(2.8E+10)	(2.7E+10)	(3.4E+10)	(7.4E+10)	(2.9E+11)
40	1.8E+06	2.8E+06	1.8E+06	1.4E+06	1.1E+06	1.8E+06	2.3E+06
(104)	(1.8E+08)	(2.8E+08)	(1.8E+08)	(1.4E+08)	(1.1E+08)	(1.8E+08)	(2.3E+08)
60	1.7E+04	2.0E+04	1.2E+04	1.4E+04	1.1E+04	2.2E+04	1.8E+04
(140)	(1.7E+06)	(2.0E+06)	(1.2E+06)	(1.4E+06)	(1.1E+06)	(2.2E+06)	(1.8E+06)

Temperature - Viscosity Curve Coefficients:
   Assuming: log(log(viscosity)) = (intercept) - (slope) x temperature

	A	B	C	D	A	C	D
Regression, R for Pa x s (R for cps)	0.9999	0.9999	0.9998	0.9998	0.9993	0.9976	0.9974
	(0.9992)	(0.9999)	(1.0000)	(0.9990)	(0.9979)	(0.9955)	(0.9944)
Intercept, log (log(Pa x s.))	1.1458	1.1529	1.1548	1.1438	1.1580	1.1611	1.2078
(log(log(cps)))	(1.2961)	(1.3024)	(1.3057)	(1.2947)	(1.3107)	(1.3126)	(1.3688)
Slope, log (log(pa x s./$^{\circ}$C))	-0.0087	-0.0086	-0.0090	-0.0088	-0.0092	-0.0088	-0.0098
(log(log(cps/$^{\circ}$F)))	(-0.0036)	(-0.0036)	(-0.0037)	(-0.0036)	(-0.0038)	(-0.0037)	(-0.0041)

Summary Statistics:	Mean	Standard Deviation	Mean Variance
Regression, R for Pa x s.	0.9991	0.0011	1.77E-07
(R for cps)	(0.9980)	(0.0022)	(7.0E-07)
Intercept, log(log(Pa x s.))	1.1606	0.0217	0.0001
(log(log(cps)))	(1.3130)	(0.0255)	(0.0001)
Slope, log(log(Pa x s./$^{\circ}$C))	-0.0090	0.0004	2.40E-08
(log(log(cps/$^{\circ}$F)))	(-0.0037)	(0.0002)	(4.6E-09)

Viscosity Studies

Table 4 shows parallel plate viscosity data on four samples of type 1 and three samples of type 3 coal-tar pitch from four different manufacturers of coal-tar pitch.

I calculated the temperature-viscosity equation using the well known linear relationship, log log viscosity = intercept - slope x temperature. This linear assumption is validated by the 0.99+ regression coefficient for each bitumen.

The variation between the calculated values for the intercepts is quite low, as is the variation between the calculated slope values. Considering the wide variation in the porosity of the glass felts, these minor variations in the coefficients of the temperature-viscosity equation are too small to consider. I used the mean values for the intercept and the slope for all of the subsequent work.

Computer Model

I constructed a mathematical model using an "ideal" roofing membrane, the temperature-viscosity curve from our viscosity study, the measured mean porosity of each glass fiber felt, and the mathematical model for generating the normal temperature of a gray surface during each hour of each day at Boston, Mass. and Miami, Fla.

The "ideal" roofing membrane selected is a gravel surfaced built-up roof composed of four plies of glass felt and coal-tar pitch, on an impervious substrate. The following masses are assumed: the gravel is 19.53 kg/m^2 (400 $lb/100 ft^2$), the top coating is 3.66 kg/m^2 (75 $lb/100 ft^2$), each felt ply is 340 g/m^2 (7 $lb/100 ft^2$), and each interply bitumen is 1.46 kg/m^2 (30 $lb/100 ft^2$).

Table 5 shows the coefficients for the thermal equations used to generate the normal temperatures on grey colored surfaces in Boston, Mass. and Miami, Fla. All of the temperatures in table 5 are in degrees F.   "A" is the mean lowest January temperature. "A'" is the lowest temperature with radiative cooling. "B" is the difference between the mean lowest June and January temperatures. "C" is the difference between the mean high and low temperatures each month.   "C'" is "C" adjusted for the temperature increase on a grey surface over the ambient air temperature. "C''" is the value of "C'" adjusted for the degree of normal cloud cover at each location. This latter value was not used in the model reported in this paper, because the average value is not representative in exponential function. Instead, I calculated the mass transfer values separately for the clear, cloudy and partly cloudy days in each month.

TABLE 5 - Coefficients for the Temperature Equations,
Boston, Mass. and Miami, Fla.

BOSTON, MA	A = 22.5	A' = 4.5	B = 42.6				
MONTH	C	C'	clear days	partly cloudy	cloudy days	C''	total days
1	14.9	92.1	9	7	15	46.0	31
2	16.1	94.2	8	7	13	48.2	28
3	16	94.0	8	8	15	46.2	31
4	18.9	99.1	7	9	14	49.6	30
5	20.5	101.8	6	11	14	50.7	31
6	19.8	100.6	7	10	13	52.1	30
7	18.8	98.8	7	12	12	52.4	31
8	18.5	98.4	9	11	11	55.9	31
9	17.9	97.3	10	8	12	55.0	30
10	17.6	96.8	11	8	12	55.9	31
11	15	92.3	8	7	15	44.6	30
12	14.7	91.8	9	7	15	45.8	31

MIAMI, FL   A = 67.2   A' = 49.2   B = 16.8

MONTH	C	C'	clear days	partly cloudy	cloudy days	C"	total days
1	16.9	109.6	9	11	11	60.2	31
2	17.6	110.8	9	9	10	62.5	28
3	16.5	108.9	10	10	11	61.2	31
4	15.4	107.0	10	12	8	64.2	30
5	11.4	100.0	10	12	9	57.1	31
6	14.1	104.7	5	14	11	50.3	30
7	13.6	103.8	3	17	11	47.1	31
8	13.6	103.8	3	17	11	47.1	31
9	11.3	99.8	4	14	12	43.8	30
10	13.6	103.8	10	11	10	58.7	31
11	15.4	107.0	12	10	8	67.3	30
12	16.6	109.0	11	9	11	62.8	31

Table 6 shows dramatically how high porosity felts settle in the roofing membrane in a warm climate. As the temperature and the porosity decline, much more time is required to settle all of the felts to the bottom of the membrane, but the mechanism is the same, if slower. I elected to use the time from the start of the test to when the bottom felt settles to reduce the lowest interply to zero. This is more for convenience than for any "failure point" concept.

TABLE 6 - Settling of High Porosity Felts in Hot Climates

Time Yr.-Mo.	Top Ctg.	Mass, Pounds per 100 Square Feet 1st Interply	2nd Interply	3rd Interply	4th Interply
0-0	75.00	30.00	30.00	30.00	30.00
0-1	93.67	31.07	31.41	31.40	7.44
0-2	110.61	31.43	32.66	20.28	0.00
0-3	130.60	31.15	33.23	0.00	0.00
0-4	146.35	30.40	18.23	0.00	0.00
0-5	156.37	29.69	8.91	0.00	0.00
0-6	185.81	9.14	0.00	0.00	0.00

Table 7 shows the results of the many model "runs" for each of the glass fiber felt types at each of the exposures selected. From a membrane stability view point (neglecting the problems associated with incompatibility), only the low porosity type IV felt from manufacturer F may be acceptable for exposure in the Boston, Mass. area.

**TABLE 7 - Porosity Through One Felt Ply and Time for
Bottom Ply to Touch Substrate**

Felt Source	ASTM Type	Flow mL/s x m^2	Time to Zero Mass, Lowest Interply, yr. Boston, Mass.	Miami, Fla.
I	D 2178 IV	13100	0.770	0.018
H	D 2178 IV	10800	1.102	0.022
G	D 2178 VI	7600	1.470	0.032
F	D 2178 VI	6990	1.563	0.035
G	D 2178 IV	2280	4.884	0.109
I	D 2178 VI	2180	5.283	0.115
F	D 2178 IV	243	20+	1.249

The porosity of all of the felts is too high for high temperature exposures such as Miami, Fla.

These data are consistent with my field observations and our laboratory work.

The graph in table 8 shows a hyperbolic relationship between the porosity of the felt and the time for the bottom felt ply to touch the substrate.

TABLE 8 - Porosity vs. Time for Lowest Felt Ply to Touch Substrate

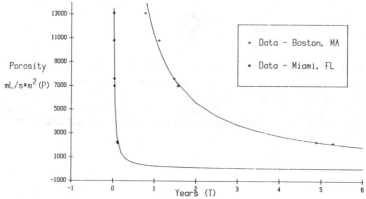

Conclusions

The following conclusions are based on my field observations, our laboratory tests, and the work reported in this paper.

All of the glass fiber felts tested have too much porosity to be used with coal-tar pitch (or, although untested, dead level asphalt) interply bitumens.

The porosity of the glass felts is highly variable, and is independent of the ASTM 2178 felt type.

Recommendations

I recommend the following, based on my field observations, laboratory tests, and the work reported in this paper:

Use only steep asphalts, D 312, Types 3 or 4, as the interply bitumen with glass felts.

The appropriate sub-committee of ASTM Committee D08 should further develop the porosity of felt test method, include a maximum value for porosity in standard D 2178 (to include a rational safety margin), and a note to warn the users not to use coal-tar pitch, or other low temperature flowing bitumens with the felts covered in this standard.

Investigate the porosity of the newer coal-tar pitch coated glass felts, and establish prudent standard values for their porosity.

Acknowledgments

The work for this paper was supported entirely by the Principals and Associates of Simpson Gumpertz & Heger Inc. The viscosity and porosity work reported would have been impossible without the diligent work of the laboratory technicians in our roofing laboratory at Arlington, Mass.

Carl G. Cash[1] and Justin Henshell[2]

STATISTICS FROM BUILT-UP ROOFING SAMPLE ANALYSES

REFERENCE: Cash, C. G. and Henshell, J., "**Statistics from Built-Up Roofing Sample Analyses,**"Roofing Research and Standards Development: 2nd Volume, ASTM STP 1088, Thomas J. Wallace and Walter J. Rossiter, Eds., American Society for Testing and Materials, Philadelphia, 1990.

ABSTRACT: This paper presents data from over 1,000 built-up roofing membrane samples analyzed by the contributing laboratories. These data were assembled in Task Force D 8.20.15, chaired by Justin Henshell, to determine the typical values that can be expected from the quantitative analysis of built-up roofing membranes.

These data are evaluated by standard statistical methods to determine the type of distribution, mean, standard deviation, range (95% confidence) and the typical values. For this paper the typical range is defined as the range that includes more than 50% of the samples.

The range of the typical values calculated are broader than the recommendations by the roofing material manufacturers, but not as broad as the recommendations of the National Roofing Contractors Association.

KEY WORDS: built-up roofing, asphalt, pitch, top coating, interply bitumen, aggregate

Introduction

ASTM Task Group D 8.20.15 has been collecting data to determine the values that can be expected from the quantitative analysis of built-up roofing samples, and has as a goal the preparation of an ASTM standard to describe the typical limits of the mass of the loose aggregate, adhered aggregate, top coating bitumen, and average interply bitumen. The proposed standard would also list the typical limits for the adhered aggregate as a percent of the applied aggregate, and the interply voids.

---

[1]Principal, Simpson Gumpertz & Heger Inc., consulting engineers, Arlington, Mass. 02174

[2]Principal, Justin Henshell Architect, Red Bank, N. J. 07701

Data contributing Task Force members are Gerald B. Curtis, Dwight F. Jennings, Ladislav Jerga, James D. Koontz, and Robert A. LaCosse. All of the test samples were cut from new roofs as a feature of the owners' quality control program for new roofs, despite the following note in D 2829, "Standard Practice for Sampling and Analysis of Built-Up Roofs": "This procedure is for the investigation of problems in existing roofs and not intended for new construction inspection."

Procedure

We entered all data into computer files, sorted each file to report data from the lowest to the highest datum, and compared the distribution of the data with their distribution in a normal probability function. We made special calculations where the distribution was skewed.

After we confirmed the fit with the normal probability function, we calculated the mean and standard deviation of each data file. We used all data; we did not exclude outlying datum (datum outside the +/- 1.96 standard deviation range).

We used the calculated mean and standard deviation to define the 95 percentile range (mean +/- 1.96 standard deviations) and the typical range (mean +/- 0.69 standard deviations for standard distributions, and 1/2 median to 1-1/2 median values for skewed distributions). We also computed the mass of the adhered aggregate as a percent of the total aggregate, and the relationship between the mass of the adhered aggregate and the mass of the bitumen flood coating.

Data in the summary tables are reported in both metric and conventional units. The original data, in appendices 1 through 11, are listed as received, in conventional units.

Correlation With the Normal Distribution

We compared the cumulative distribution of data in each file to the cumulative values in a normal distribution. Table 1 shows the least squares linear regression coefficient for each file. Perfect correlation would result in a regression value of 1.0.

The lowest regression values we obtained is 0.964 for the voids in interply coal-tar pitch and 0.976 for the voids in interply asphalt where many samples showed zero percent voids, skewing the distribution. Aside from data on voids, these values show data in each file approximates a normal distribution.

**TABLE 1 - Linear Regression Coefficient, Cumulative
Data Distribution vs. Cumulative Normal Distribution**

	Asphalt	Pitch
Loose Aggregate	0.995	0.999
Adhered Aggregate	0.995	0.998
Top (Flood) Coating	0.986	0.997
Average Interply Bitumen	0.998	0.999
Percent Voids	0.976	0.964
% Adhered Aggregate	0.999	0.999

Summary Statistics

Table 2 lists the mean, standard deviation and the number of samples in each file. The difference between the data for coal-tar pitch and asphalt membranes is statistically insignificant.

TABLE 2 - Summary Statistics, All Data

Bitumen Type: Units	Asphalt		Pitch	
	kg/m^2	lb/100ft^2	kg/m^2	lb/100ft^2
Loose Aggregate				
Mean	9.7	198	8.7	178
Standard Deviation	6.2	126	6.0	123
Number of Samples		930		168
Adhered Aggregate				
Mean	13.3	272	13.2	271
Standard Deviation	5.8	119	4.7	97
Number of Samples		1183		223
Top (Flood) Coating				
Mean	3.9	79	3.9	81
Standard Deviation	2.4	49	1.5	30
Number of Samples		1254		225
Average Interply Bitumen				
Mean	1.2	26	1.4	29
Standard Deviation	0.3	6	0.3	5
Number of Samples		1254		225
Interply Voids				
Mean % of Sample Area		2.7		2.3
Standard Deviation		5.0		5.2
Median (mid-value)		1.4		0.2
Number of Samples		1138		169
% Adhered Aggregate				
Mean % Applied Aggregate		59		56
Standard Deviation		19		16
Number of Samples		905		142

Interply Bitumen in Organic and Glass Felt Membranes

We compared the average interply bitumen data for samples composed of organic and glass fiber felts. The results are shown in Table 3.

TABLE 3 - Comparative Statistics, Asphalt and Coal-Tar Pitch with Organic and Glass Felts

Bitumen Felt	Pitch Organic	Asphalt Organic	Pitch Glass	Asphalt Glass
Average Interply Bitumen,	g/m^2		(lb./100 ft^2)	
Mean (X)	1320(27)	1360(28)	1540(32)	1400(29)
Standard Deviation	232(4.8)	213(4.5)	271(5.6)	301(6.2)
Mean Variance (V)	11(0.23)	20(0.40)	34(0.69)	4(0.07)
Number of Samples	100	47	45	513
PITCH VS. ASPHALT	Organic		Glass	
Effective Frequency	100		54	
s Student's $t_{(0.75)}$ $u = t \times (V_{(a)} + V_{(b)})^{0.5}$ $X_{(a)} - X_{(b)} =$	1.98 85(1.8) 8(0.4)		2.01 85(1.8) 137(2.8)	
ORGANIC VS. GLASS	Pitch		Asphalt	
Effective Frequency	76		65	
Student's $t_{(0.75)}$ $u = t \times (V_{(a)} + V_{(b)})^{0.5}$ $X_{(a)} - X_{(b)} =$	1.99 93(1.9) 204(4.2)		1.99 67(1.4) 48(1.0)	

These data confirm that the overall difference between coal-tar pitch and asphalt interplies is not statistically different at the 95% confidence level, but that the mass of interply coal-tar pitch used with glass felts is statistically significantly higher than the mass of the interply asphalt or the interply pitch used with organic felts.

Built-Up Roofing Components, Range at 95% Confidence

Table 4 shows the range of data for individual samples with a 95% probability (t = 0.975). The ranges for loose aggregate and top bitumen coating are very wide. The interply bitumen ranges are relatively small.

TABLE 4 - Range of Analytical Values

kg/m^2(lb./100 ft^2) unless otherwise noted

Bitumen Type	Asphalt	Pitch
Loose Aggregate	0(0) to 21.7(445)	0(0) to 20.5(419)
Adhered Aggregate	1.9(39) to 24.7(505)	3.9(81) to 22.5(461)
Top (Flood) Coating	0(0) to 8.5(175)	1.1(22) to 6.8(140)
Average Interply Bitumen with Organic Felts	0.9(19) to 1.8(37)	0.9(18) to 1.8(37)
Average Interply Bitumen with Glass Felts	0.8(17) to 2.0(41)	1.0(20) to 2.1(43)
Interply Voids, Percent of Sample Area	0 to 16	0 to 18
Adhered Aggregate, % of Applied Aggregate	22 to 96	24 to 88

Built-Up Roofing Components, Typical Values

Table 5 shows the typical values for each of the elements of the built-up roofing membrane measured.

TABLE 5 - Typical* Analytical Values for Built-Up Roofing Membranes

kg/m^2(lb./100 ft^2) unless otherwise noted Bitumen Type	Asphalt	Pitch
Loose Aggregate	5.5(112) to 13.8(284)	4.6(94) to 12.8(262)
Adhered Aggregate	9.3(191) to 17.2(353)	10.0(205) to 16.4(337)
Top (Flood) Coating	2.2(46) to 5.5(112)	3.0(61) to 5.0(101)
Average Interply Bitumen with Organic Felts	1.2(25) to 1.5(31)	1.2(24) to 1.5(31)
Average Interply Bitumen with Glass Felts	1.2(25) to 1.6(33)	1.4(28) to 1.7(35)
Interply Voids, Percent of Sample Area	0 to 2	0 to 2
Adhered Aggregate, % of Applied Aggregate	46 to 72	45 to 67

*Range of 50% of Samples Tested

Adhered Aggregate vs. Top (Flood) Coating Mass

We calculated the linear regression coefficients for the mass of the adhered aggregate vs. the mass of the top (flood) coating. The regression coefficient, R, is a modest 0.620 for asphalt and 0.678 for coal-tar pitch. The linear equations are quite similar:

- for asphalt:
  Y (Adhered Aggregate) = 80 + 2.26 X (Top Coating)

- for coal-tar pitch:
  Y (Adhered Aggregate) = 87 + 2.25 X (Top Coating)

Manufacturers' and NRCA Recommendations vs.Typical Values

Data in Table 6 compare the roofing material manufacturers' and the National Roofing Contractors Association (NRCA) recommendations for some of the values covered in this report.

TABLE 6 - Recommendations and Typical Values Compared, lb./100 ft.^{2}

	Manufacturer's Recommendations	NRCA Recommendations	Typical Values
Top Pitch Coating	70 minimum	56-94	61-101
Top Asphalt Coating	51-70	45-75	46-112
Avg. Interply Pitch	20 minimum	19-31	26-32
Avg. Interply Asphalt	20-30	15-25	22-30

We placed these values in the same table, but they are not strictly comparable since the manufacturers' recommendations imply a range for a value for any sample, as does the typical values here presented.

The much wider range of the NRCA recommendations is for the average of all the samples for one job. The current NRCA recommendations are far too broad, based on this data.

Summary

The typical values suggested in Table 5 are both realistic and attainable. They have a wider range than those recommended by the manufacturers, but much narrower than the NRCA recommendations. We hope that a new ASTM standard, setting forth these values will be approved

Acknowledgements

This paper is dedicated to the many laboratory technicians whose labor in obtaining these data made this paper possible.

Appendix 1 Loose Aggregate on Asphalt Built-up
Roofing Membranes, lbs per 100 square feet

1	57	98	125	156	173	193	219	250	292	411
3	58	98	125	156	174	194	219	251	295	413
3	58	99	125	156	174	194	219	251	296	413
3	58	99	125	156	174	194	220	251	296	414
3	58	99	126	157	174	194	220	252	296	416
4	58	99	127	158	174	194	220	252	296	425
4	59	100	127	158	174	194	221	252	298	425
4	59	100	127	158	174	195	222	252	300	431
4	59	101	127	158	174	195	222	252	301	436
4	60	101	127	159	174	195	222	253	303	451
5	60	102	128	159	174	196	223	253	305	453
7	61	102	128	159	175	196	223	254	306	454
7	61	102	128	159	178	196	223	254	306	456
8	62	102	129	159	178	196	223	254	307	469
8	63	103	129	159	179	196	224	254	308	469
10	63	103	130	159	179	197	224	255	308	472
10	65	103	130	160	179	197	224	256	308	475
11	65	103	131	160	180	197	224	257	308	477
13	65	103	131	160	180	197	225	258	310	488
13	66	103	132	160	180	197	226	258	310	488
13	68	103	132	160	181	198	226	258	310	496
13	68	104	132	160	181	198	226	258	311	497
14	68	104	132	160	181	198	226	259	313	500
14	69	105	132	161	181	198	227	260	313	503
14	69	106	133	161	181	198	227	261	313	504
15	70	106	133	161	181	199	227	261	314	511
16	70	106	133	161	181	199	227	262	315	512
16	70	107	133	161	181	199	228	262	316	512
18	70	107	133	161	181	199	228	263	317	516
18	71	108	135	161	182	199	228	263	319	517
18	71	108	136	162	182	200	229	264	319	536
19	72	109	137	162	182	200	230	266	320	544
19	72	109	137	162	182	200	230	266	321	552
21	73	109	137	162	182	200	231	266	324	561
21	75	109	138	162	183	200	231	267	326	579
22	75	110	138	162	183	201	232	268	331	612
23	76	110	138	162	184	201	232	268	331	614
24	76	110	139	163	184	201	233	268	333	620
26	77	110	140	163	184	201	233	270	334	632
26	77	110	140	163	184	201	233	271	335	654
28	77	111	140	163	185	203	233	271	336	657
28	77	111	140	163	185	203	234	271	338	661
28	78	112	140	164	186	203	234	271	338	661
28	79	112	141	164	186	203	234	272	341	673
30	80	112	141	164	186	204	235	272	341	682
30	80	113	141	164	186	204	235	272	342	683
31	82	113	142	164	186	204	236	273	345	701
33	82	113	142	164	187	204	236	273	346	708
34	82	113	142	165	187	204	238	274	347	710
34	83	113	143	165	187	205	238	275	347	715
34	83	113	143	166	187	205	238	275	348	737
37	83	114	144	166	187	206	239	276	350	766
39	84	114	144	166	187	206	239	277	352	783

Appendix 1 Loose Aggregate on Asphalt (continued)

40	84	114	145	166	187	206	239	277	353	873
40	84	115	145	167	188	206	240	278	353	
42	84	116	145	167	188	206	240	278	354	
42	84	116	146	167	188	207	240	279	354	
42	85	116	146	167	188	207	241	280	354	
43	85	117	147	168	189	207	241	281	356	
43	86	117	147	168	189	208	242	282	356	
43	87	118	147	168	189	208	242	282	358	
45	87	118	147	168	189	208	242	283	359	
45	87	118	147	168	189	209	242	284	359	
47	88	119	148	168	190	209	243	284	360	
48	89	120	149	168	190	210	243	285	360	
49	89	120	150	169	190	210	243	285	362	
50	90	120	150	169	190	211	243	285	363	
50	91	120	151	169	190	212	243	285	364	
50	91	120	151	169	190	212	244	285	368	
50	92	120	151	169	190	212	244	286	369	
50	93	120	152	169	190	212	244	286	370	
50	94	121	152	170	191	213	244	286	371	
50	94	121	152	171	191	213	245	286	373	
51	94	121	152	171	191	214	245	287	374	
51	94	121	152	171	191	214	245	288	380	
52	95	121	152	171	191	216	245	288	380	
53	95	122	153	171	191	216	245	288	384	
53	95	122	153	171	191	217	247	289	384	
53	96	122	154	171	191	217	247	290	390	
54	97	123	154	172	192	217	247	290	390	
55	97	123	154	173	192	217	248	291	391	
55	97	124	155	173	192	217	248	291	393	
55	97	124	155	173	192	218	248	292	395	
56	97	124	155	173	193	218	248	292	407	
56	97	125	155	173	193	218	248	292	408	
57	98	125	155	173	193	219	249	292	410	

Appendix 2 Loose Aggregate on Pitch Built-up Roofing Membranes, lbs per 100 square feet

22	104	138	159	174	195	221	260	313	395
30	108	141	159	174	195	221	264	320	396
37	109	142	160	174	199	228	268	321	434
38	110	142	162	177	202	233	269	325	452
43	111	142	164	179	203	234	276	329	476
47	114	143	166	182	206	237	281	345	532
59	121	152	166	183	206	332	285	349	696
59	121	155	167	183	207	240	290	350	
61	124	155	167	185	213	240	295	351	
71	125	156	168	186	215	241	295	355	
76	129	156	168	187	215	242	303	363	
77	135	157	172	188	216	243	308	365	
83	135	158	173	191	218	246	309	377	
94	137	158	173	193	220	249	311	379	
98	138	159	173	193	221	257	312	379	

Appendix 3   Adhered Aggregate on Asphalt Built-up
Roofing Membranes, lbs per 100 square feet

18	151	175	194	210	227	245	271	298	331	374	436
27	151	175	194	211	227	246	272	298	331	375	436
41	151	175	194	211	227	246	272	298	332	376	437
41	151	175	194	211	227	246	272	298	332	376	438
49	152	175	195	211	228	247	272	298	333	376	441
50	152	175	195	211	228	247	273	299	333	376	445
52	152	176	195	211	228	247	273	299	334	376	445
59	152	176	195	212	228	247	273	299	334	376	446
60	153	176	195	212	229	247	273	299	335	377	447
63	153	176	195	212	229	247	273	299	336	377	451
64	154	177	195	212	229	248	274	299	336	377	451
66	154	177	195	212	229	248	275	300	336	378	452
67	154	177	195	212	229	248	275	300	337	378	452
69	154	177	196	213	229	249	275	300	337	378	453
83	154	177	196	213	229	249	276	300	338	378	453
84	155	177	196	213	229	249	276	300	338	378	453
84	155	178	196	213	229	249	276	300	338	378	453
84	155	178	196	214	230	249	276	301	338	379	454
84	155	178	197	214	230	249	276	301	339	379	456
86	155	178	197	214	230	249	276	301	340	379	457
87	155	179	197	214	230	250	277	301	340	380	458
88	156	179	197	214	230	250	277	302	340	381	459
91	156	179	197	214	230	251	277	303	340	381	462
91	157	179	197	214	230	252	278	304	341	383	465
93	157	180	198	215	230	253	278	305	342	383	468
94	157	180	198	215	231	253	278	305	343	383	468
96	157	180	198	215	232	253	278	305	343	383	471
98	158	180	198	215	232	253	279	306	343	383	474
98	158	181	198	215	233	253	279	306	343	384	476
99	158	181	198	215	233	253	279	306	343	386	477
102	159	181	198	215	233	253	279	306	344	386	478
103	159	181	198	215	234	254	279	306	344	386	478
104	159	181	198	216	234	254	280	306	344	387	484
105	159	181	199	216	234	254	280	306	346	387	485
107	159	181	199	216	234	255	280	306	347	387	489
107	159	181	199	217	234	255	280	307	347	388	490
107	159	181	200	217	234	255	280	307	347	388	491
108	159	181	200	217	235	256	280	308	347	390	491
110	160	182	200	217	235	256	280	308	348	392	495
110	160	182	200	217	235	256	280	308	348	394	496
111	160	182	200	218	235	257	280	308	348	394	496
112	160	182	200	218	235	257	281	309	349	394	500
112	161	182	201	218	235	257	281	309	349	394	503
114	161	183	201	218	235	257	281	309	350	395	507
114	161	183	201	218	235	258	282	309	350	396	514
116	161	183	201	218	235	258	283	310	350	396	514
117	162	183	201	218	235	258	283	310	350	396	516
119	162	183	201	219	236	258	283	310	350	397	516
120	162	184	201	219	236	258	283	310	351	397	519
123	162	184	201	219	237	259	283	311	351	397	520
124	162	184	201	219	237	259	283	311	352	398	523
126	163	184	201	219	237	259	284	311	353	398	531
126	163	184	202	219	237	259	285	311	353	400	533

Appendix 3 Adhered Aggregate on Asphalt (continued)

126	163	185	202	219	237	259	285	311	353	400	534
129	163	185	202	220	238	259	285	311	353	401	536
130	163	185	202	220	238	260	286	314	354	402	539
131	164	185	202	220	238	260	286	314	355	404	541
132	164	185	202	220	238	260	286	314	355	408	541
132	165	186	202	221	239	260	286	315	355	409	546
132	165	186	203	221	239	260	287	315	355	410	547
132	166	186	203	221	239	260	288	315	356	410	548
133	166	187	203	221	239	260	288	315	356	410	550
134	166	187	203	221	239	261	288	316	357	410	550
135	167	187	203	221	239	261	289	316	357	410	560
135	167	187	203	222	239	261	289	317	357	411	560
135	167	187	203	222	239	261	289	317	357	412	569
136	167	187	204	222	240	261	289	317	357	412	572
136	167	187	204	222	240	261	289	317	357	412	574
138	168	188	204	222	240	261	289	318	360	413	587
138	168	188	204	222	240	261	289	318	360	414	589
138	168	188	205	222	240	262	290	319	360	415	591
138	168	188	205	223	240	262	290	320	361	415	591
138	168	189	205	223	241	262	290	320	361	417	598
139	168	189	206	223	241	262	291	320	362	417	598
140	169	189	206	223	241	263	292	321	362	418	602
140	170	189	206	223	241	263	292	321	363	419	611
140	170	189	206	223	241	264	292	322	363	421	612
140	170	189	206	223	242	264	292	323	363	421	613
140	170	189	207	223	243	264	293	323	364	421	640
140	172	189	207	223	243	264	293	323	365	422	650
140	117	190	207	224	243	264	293	324	365	422	660
140	117	190	207	224	243	265	293	324	365	422	689
143	172	190	208	224	243	266	293	324	367	423	709
143	172	190	208	224	243	266	294	324	367	423	717
143	172	190	208	224	243	266	294	325	367	424	719
144	172	190	208	224	243	266	295	325	368	424	721
145	172	190	208	224	243	266	295	325	368	427	837
145	172	191	208	224	243	266	295	326	369	427	841
146	173	191	208	225	244	267	295	326	369	428	901
146	173	191	209	225	244	267	296	326	369	429	912
146	173	191	209	225	244	267	296	326	369	429	924
146	173	191	209	225	244	267	296	327	369	430	958
147	174	191	209	226	244	268	296	327	369	430	980
149	174	192	209	226	245	269	296	328	370	430	
149	174	192	210	226	245	269	296	328	370	431	
150	174	193	210	226	245	269	296	328	371	432	
150	174	193	210	226	245	269	297	329	371	432	
150	175	193	210	226	245	269	297	330	373	432	
151	175	194	210	227	245	270	298	331	373	434	

Appendix 4 Adhered Aggregate on Pitch Built-up Roofing Membranes, lbs per 100 square feet

55	161	178	202	224	241	258	278	312	344	379	429
88	165	179	202	225	241	259	278	314	346	379	430
92	166	179	202	228	242	259	281	314	348	384	431

Appendix 4   Adhered Aggergate on Pitch (continued)

99	167	179	205	229	242	261	282	316	348	388	439
102	167	184	207	230	242	261	286	317	350	399	446
114	167	185	207	230	243	261	287	318	351	401	453
119	167	186	208	230	243	263	288	319	351	407	459
119	167	187	209	231	243	264	290	319	351	411	493
127	168	187	209	231	244	264	291	323	358	413	499
134	169	190	211	231	244	265	292	324	359	419	509
135	170	191	215	233	244	267	293	327	360	420	512
136	171	192	216	233	244	268	293	328	360	421	527
146	172	193	216	234	246	272	293	328	360	422	546
147	172	193	217	237	247	272	297	328	360	423	574
148	173	196	218	240	249	274	297	329	362	424	
152	173	200	219	240	250	274	297	331	367	425	
158	175	200	219	240	250	274	303	333	370	425	
158	176	200	220	241	256	276	306	338	377	426	
160	178	201	221	241	256	277	312	342	378	429	

Appendix 5   Percent of Adhered Aggregate on Asphalt Built-up Roofing Membranes.

9	34	40	46	50	54	58	63	68	73	80	91
9	34	41	46	50	54	58	63	68	73	81	92
9	34	41	46	50	54	58	63	68	73	81	92
9	34	41	46	50	54	58	63	68	73	81	92
10	34	41	46	50	54	58	63	68	73	81	92
12	34	41	46	50	54	58	63	68	73	81	92
13	34	41	46	50	54	59	63	68	74	81	92
14	34	41	46	50	54	59	63	68	74	81	92
14	35	41	46	50	54	59	63	68	74	81	92
15	35	41	46	50	54	59	63	68	74	82	92
15	35	41	46	50	54	59	63	68	74	82	92
18	35	42	46	50	54	59	63	68	74	82	93
18	35	42	46	50	54	59	63	68	74	82	93
18	35	42	46	50	54	59	63	68	74	82	93
18	35	42	46	50	54	59	63	68	74	82	93
19	35	42	46	50	54	59	63	68	74	82	93
19	35	42	46	50	54	59	63	68	74	82	93
19	35	42	46	50	55	59	63	68	74	82	93
19	35	42	46	50	55	59	63	69	75	83	94
20	36	42	47	50	55	59	63	69	75	83	94
20	36	42	47	50	55	59	63	69	75	83	94
20	36	42	47	51	55	59	64	69	75	83	94
21	36	42	47	51	55	59	64	69	75	83	94
22	36	42	47	51	55	59	64	69	75	84	94
23	36	42	47	51	55	60	64	69	75	84	95
23	36	42	47	51	55	60	64	69	75	84	95
24	36	42	47	51	55	60	64	69	75	84	95
24	36	43	47	51	55	60	64	69	75	84	95
24	37	43	47	51	55	60	64	69	75	84	95
25	37	43	47	51	55	60	65	69	75	84	95
25	37	43	47	51	55	60	65	69	75	84	96
25	37	43	47	51	55	60	65	69	75	84	96
25	37	43	47	51	55	60	65	69	75	85	96

Appendix 5 Percent of Adhered Aggregate on Asphalt (continued)

26	37	43	47	51	55	60	65	69	75	85	97
26	37	43	47	51	55	60	65	69	76	85	97
27	38	43	47	51	55	60	65	69	76	85	97
27	38	43	47	51	55	60	65	70	76	85	97
27	38	43	48	51	56	60	65	70	76	85	97
28	38	43	48	51	56	61	65	70	76	85	97
28	38	43	48	51	56	61	65	70	76	85	97
28	38	43	48	51	56	61	65	70	76	85	97
28	38	43	48	52	56	61	65	70	77	85	97
29	38	43	48	52	56	61	65	70	77	86	97
29	38	43	48	52	56	61	65	70	77	86	98
29	38	43	48	52	56	61	65	70	77	86	98
29	39	43	48	52	56	61	65	70	77	86	98
30	39	43	48	52	56	61	65	70	77	87	98
30	39	43	48	52	56	61	65	70	77	87	98
30	39	44	48	52	56	61	65	70	77	87	99
30	39	44	48	52	57	61	66	71	77	87	99
30	39	44	48	52	57	61	66	71	77	87	99
30	39	44	48	52	57	61	66	71	77	87	99
31	39	44	48	52	57	61	66	71	77	87	99
31	39	44	48	52	57	61	66	71	77	87	99
31	39	44	49	52	57	61	66	71	78	88	99
31	40	44	49	52	57	61	66	71	78	88	99
31	40	44	49	52	57	62	66	71	78	88	99
32	40	44	49	53	57	62	66	71	78	88	100
32	40	44	49	53	57	62	66	71	78	88	
32	40	44	49	53	57	62	66	71	78	88	
32	40	45	49	53	57	62	66	71	78	88	
32	40	45	49	53	57	62	66	71	78	88	
32	40	45	49	53	57	62	66	71	78	88	
32	40	45	49	53	57	62	67	71	79	88	
32	40	45	49	53	57	62	67	71	79	88	
32	40	45	49	53	57	62	67	71	79	88	
32	40	45	49	53	57	62	67	72	79	89	
32	40	45	49	53	57	62	67	72	79	89	
33	40	45	49	53	57	62	67	72	79	89	
33	40	45	49	53	57	62	67	72	79	89	
33	40	45	49	53	58	62	67	72	79	90	
33	40	45	49	53	58	62	67	72	79	90	
33	40	45	49	53	58	62	67	72	80	90	
33	40	45	49	53	58	62	67	72	80	90	
33	40	45	49	54	58	62	67	73	80	91	
34	40	45	49	54	58	63	67	73	80	91	
34	40	46	49	54	58	63	68	73	80	91	

Appendix 6 Percent of Adhered Aggregate on Pitch Built-up Roofing Membranes

16	32	39	44	50	54	58	61	65	67	71	80
22	32	39	46	50	54	58	61	65	67	72	83
25	32	39	46	50	54	58	62	65	67	73	87
29	32	40	46	50	54	59	62	65	67	73	87
29	34	40	46	51	55	59	62	65	69	73	88

Appendix 6 Percent of Adhered Aggregate (continued)

29	34	41	46	51	55	59	63	65	69	74	88
29	35	41	48	51	56	59	63	65	69	75	90
30	36	42	49	51	56	59	63	66	69	77	90
31	37	42	49	52	57	59	63	66	70	77	93
31	37	42	49	53	57	59	63	66	71	77	94
32	37	42	50	54	58	60	64	67	71	78	
32	39	43	50	54	58	61	64	67	71	79	

Appendix 7 Top Coating on Asphalt Built-up Roofing Membranes, lbs per 100 square feet

5	44	51	55	59	64	68	74	83	94	108	131
8	44	51	55	59	64	68	74	83	95	108	131
11	44	51	56	60	64	68	74	83	95	109	131
13	44	51	56	60	64	68	74	83	95	109	132
14	44	51	56	60	64	68	74	83	95	109	132
14	44	51	56	60	64	68	74	83	95	109	132
17	45	51	56	60	65	69	74	84	96	109	133
17	45	51	56	60	65	69	74	84	96	109	133
18	45	51	56	60	65	69	74	84	96	109	134
19	45	51	56	60	65	69	74	84	96	109	136
20	45	51	56	60	65	69	74	84	96	109	136
20	45	51	56	60	65	69	75	84	96	110	137
21	45	51	56	60	65	69	75	84	96	110	137
26	45	51	56	60	65	69	75	84	96	110	137
26	45	51	56	60	65	69	75	85	96	110	138
28	45	51	56	60	65	69	75	85	96	110	138
28	45	52	56	60	65	69	75	85	97	110	138
29	45	52	56	60	65	69	75	85	97	111	139
29	45	52	56	60	65	69	75	85	97	111	139
30	45	52	57	60	65	69	75	85	97	111	139
30	45	52	57	60	65	69	75	85	97	111	139
31	45	52	57	60	65	69	75	86	97	111	141
31	45	52	57	60	65	69	76	86	97	112	141
31	46	52	57	60	65	69	76	86	98	112	141
31	46	52	57	61	65	69	76	86	98	112	142
32	46	52	57	61	65	69	76	86	98	112	143
32	46	52	57	61	65	69	76	86	98	113	143
32	46	52	57	61	65	69	76	86	98	113	143
32	46	52	57	61	65	69	76	86	98	113	143
32	46	52	57	61	66	69	76	86	98	113	144
32	46	52	57	61	66	69	76	86	98	113	145
32	46	52	57	61	66	69	76	87	98	113	145
32	46	52	57	61	66	70	77	87	98	113	145
33	46	52	57	61	66	70	77	87	99	113	146
33	46	52	57	61	66	70	77	87	99	113	146
34	46	52	57	61	66	70	77	87	99	114	146
34	46	52	57	61	66	70	77	87	99	114	147
34	47	52	57	61	66	70	77	87	99	114	147
35	47	52	57	61	66	70	77	87	99	114	147
35	47	52	57	61	66	70	77	87	99	114	147
35	47	53	57	61	66	70	77	87	99	114	149

Appendix 7 Top Coating on Asphalt (continued)

35	47	53	57	61	66	70	77	87	99	114	149
36	47	53	57	61	66	70	77	87	99	115	149
36	47	53	57	61	66	70	78	88	100	115	150
36	47	53	57	62	66	70	78	88	100	115	150
37	47	53	57	62	66	70	78	88	100	115	150
37	47	53	57	62	66	70	78	88	100	115	151
37	47	53	57	62	66	70	78	88	100	116	151
37	47	53	57	62	66	71	78	88	100	117	152
37	47	53	57	62	66	71	78	88	100	117	152
37	48	53	57	62	66	71	78	88	100	117	153
37	48	53	57	62	66	71	78	88	100	117	154
37	48	53	57	62	66	71	78	89	101	117	154
38	48	53	57	62	66	71	78	89	101	117	154
38	48	53	58	62	66	71	78	89	101	117	155
38	48	53	58	62	66	71	78	89	101	118	155
38	48	53	58	62	66	71	79	89	101	118	155
39	48	53	58	62	66	71	79	89	101	118	155
39	48	53	58	62	66	71	79	89	101	118	156
39	48	53	58	62	66	71	79	89	101	118	157
39	48	53	58	62	66	71	79	89	101	118	159
39	48	53	58	62	66	71	79	89	102	119	160
39	48	53	58	62	67	71	79	89	102	119	163
39	48	53	58	62	67	71	79	89	102	119	165
39	48	53	58	62	67	71	79	89	102	119	166
39	48	53	58	62	67	71	79	89	102	120	166
39	49	54	58	63	67	71	79	89	102	120	167
40	49	54	58	63	67	71	79	90	102	120	167
40	49	54	58	63	67	72	80	90	103	121	169
40	49	54	58	63	67	72	80	90	103	121	172
40	49	54	58	63	67	72	80	90	103	121	177
40	49	54	58	63	67	72	80	90	103	121	177
40	49	54	58	63	67	72	80	90	103	122	178
40	49	54	58	63	67	72	80	90	103	122	179
40	49	54	58	63	67	72	80	90	104	122	181
40	49	54	58	63	67	72	80	90	104	122	184
40	49	54	58	63	67	72	81	91	104	123	184
40	49	54	58	63	67	72	81	91	104	124	189
40	49	54	58	63	67	72	81	91	104	124	190
41	49	55	58	63	67	72	81	91	104	124	190
41	49	55	58	63	68	72	81	92	105	124	192
41	49	55	58	63	68	72	81	92	105	124	193
41	49	55	58	63	68	72	81	92	105	124	196
41	49	55	58	63	68	73	81	92	105	125	206
41	49	55	59	63	68	73	81	92	106	126	209
42	50	55	59	63	68	73	81	92	106	126	215
42	50	55	59	63	68	73	81	92	106	126	219
42	50	55	59	63	68	73	82	92	106	127	224
42	50	55	59	63	68	73	82	92	106	127	229
42	50	55	59	63	68	73	82	93	106	127	238
42	50	55	59	63	68	73	82	93	106	127	239
42	50	55	59	64	68	73	82	93	107	127	242
42	50	55	59	64	68	73	82	93	107	128	263
43	50	55	59	64	68	73	82	93	107	128	266
43	50	55	59	64	68	73	82	93	107	128	272

Appendix 7 Top Coating on Asphalt (continued)

43	50	55	59	64	68	73	82	93	107	128	294
43	50	55	59	64	68	73	82	93	107	129	364
43	50	55	59	64	68	74	83	94	108	129	
43	50	55	59	64	68	74	83	94	108	129	
43	50	55	59	64	68	74	83	94	108	129	
43	50	55	59	64	68	74	83	94	108	130	
43	50	55	59	64	68	74	83	94	108	130	
43	51	55	59	64	68	74	83	94	108	130	
44	51	55	59	64	68	74	83	94	108	130	

Appendix 8 Top Coating on Pitch Built-up
Roofing Membranes, lbs per 100 square feet

25	47	56	62	67	70	78	83	89	102	119	172
27	47	57	62	67	71	78	84	89	103	119	172
28	48	58	62	67	71	79	84	89	105	119	172
28	48	58	63	67	71	79	84	91	105	124	174
33	49	58	64	67	72	79	84	92	106	125	185
34	50	59	64	67	72	79	84	93	107	126	
35	52	59	64	68	72	81	85	93	107	127	
37	52	59	64	68	73	81	85	93	108	127	
40	53	60	64	68	73	82	85	94	109	129	
40	53	60	64	68	75	82	85	95	110	130	
41	54	60	65	68	75	82	85	95	110	137	
41	54	60	65	69	75	82	85	96	111	138	
42	54	60	65	69	76	82	85	96	111	139	
43	54	60	65	69	76	82	86	97	113	141	
43	55	60	65	70	76	82	86	97	114	142	
44	55	61	65	70	77	83	86	98	114	147	
44	55	61	65	70	77	83	87	98	117	148	
46	55	61	65	70	77	83	87	100	117	151	
47	55	61	66	70	78	83	87	101	118	158	
47	56	61	66	70	78	83	88	101	119	158	

Appendix 9 Average Interply Asphalt in Asphalt Built-up
Roofing Membranes, lbs per 100 square feet

13	19	20	22	23	24	26	27	28	30	32	36
14	19	20	22	23	24	26	27	28	30	32	36
15	19	20	22	23	24	26	27	28	30	32	36
15	19	20	22	23	24	26	27	28	30	32	36
15	19	20	22	23	24	26	27	28	30	32	36
15	19	20	22	23	24	26	27	28	30	32	36
15	19	21	22	23	24	26	27	28	30	32	36
15	19	21	22	23	24	26	27	28	30	32	36
15	19	21	22	23	24	26	27	28	30	32	36
15	19	21	22	23	24	26	27	28	30	32	37
15	19	21	22	23	24	26	27	28	30	32	37
15	19	21	22	23	25	26	27	28	30	32	37
15	19	21	22	23	25	26	27	28	30	32	37
15	19	21	22	23	25	26	27	28	30	32	37
15	19	21	22	23	25	26	27	28	30	32	37
15	19	21	22	23	25	26	27	28	30	32	37

Appendix 9 Average Interply Asphalt (continued)

16	19	21	22	23	25	26	27	28	30	32	37
16	19	21	22	23	25	26	27	28	30	32	37
16	19	21	22	23	25	26	27	28	30	33	37
16	19	21	22	23	25	26	27	28	30	33	37
16	19	21	22	23	25	26	27	28	30	33	37
16	19	21	22	23	25	26	27	28	30	33	37
16	19	21	22	23	25	26	27	28	30	33	37
16	19	21	22	23	25	26	27	28	30	33	37
16	19	21	22	23	25	26	27	28	30	33	37
16	19	21	22	23	25	26	27	28	30	33	37
16	19	21	22	23	25	26	27	28	30	33	38
16	19	21	22	23	25	26	27	28	30	33	38
16	19	21	22	23	25	26	27	28	30	33	38
16	19	21	22	23	25	26	27	29	30	33	38
16	19	21	22	23	25	26	27	29	30	33	38
17	19	21	22	24	25	26	27	29	30	33	38
17	19	21	22	24	25	26	27	29	30	33	38
17	19	21	22	24	25	26	27	29	31	33	38
17	19	21	22	24	25	26	27	29	31	33	38
17	19	21	22	24	25	26	27	29	31	33	38
17	19	21	22	24	25	26	27	29	31	33	38
17	19	21	22	24	25	26	27	29	31	33	38
17	19	21	22	24	25	26	27	29	31	34	38
17	19	21	22	24	25	26	27	29	31	34	38
17	19	21	22	24	25	26	27	29	31	34	39
17	19	21	22	24	25	26	27	29	31	34	39
17	19	21	22	24	25	26	27	29	31	34	39
17	20	21	22	24	25	26	27	29	31	34	39
17	20	21	22	24	25	26	27	29	31	34	39
17	20	21	22	24	25	26	27	29	31	34	39
17	20	21	22	24	25	26	27	29	31	34	39
17	20	21	22	24	25	26	27	29	31	34	39
17	20	21	22	24	25	26	27	29	31	34	39
17	20	21	22	24	25	26	27	29	31	34	39
17	20	21	22	24	25	26	27	29	31	34	39
17	20	21	22	24	25	26	27	29	31	34	39
17	20	21	22	24	25	26	27	29	31	34	39
17	20	21	22	24	25	26	28	29	31	34	39
17	20	21	23	24	25	26	28	29	31	34	40
17	20	21	23	24	25	26	28	29	31	34	40
18	20	21	23	24	25	26	28	29	31	34	40
18	20	21	23	24	25	26	28	29	31	34	40
18	20	21	23	24	25	26	28	29	31	34	40
18	20	21	23	24	25	26	28	29	31	34	40
18	20	21	23	24	25	26	28	29	31	34	40
18	20	21	23	24	25	26	28	29	31	34	41
18	20	21	23	24	25	26	28	29	31	34	41
18	20	21	23	24	25	26	28	29	31	34	41

Appendix 9 Average Interply Asphalt (continued)

18	20	21	23	24	25	26	28	29	31	34	41
18	20	21	23	24	25	26	28	29	31	34	41
18	20	21	23	24	25	26	28	29	31	35	41
18	20	21	23	24	25	26	28	29	31	35	41
18	20	21	23	24	25	26	28	29	31	35	41
18	20	21	23	24	25	26	28	29	31	35	41
18	20	21	23	24	25	26	28	29	31	35	42
18	20	21	23	24	25	26	28	29	31	35	42
18	20	21	23	24	25	26	28	29	31	35	42
18	20	21	23	24	25	26	28	29	31	35	43
18	20	21	23	24	25	26	28	30	31	35	43
18	20	21	23	24	25	26	28	30	31	35	43
18	20	21	23	24	25	26	28	30	31	35	43
18	20	21	23	24	25	27	28	30	31	35	43
18	20	21	23	24	25	27	28	30	31	35	43
18	20	21	23	24	25	27	28	30	31	35	44
18	20	21	23	24	25	27	28	30	31	35	44
18	20	21	23	24	25	27	28	30	31	35	44
18	20	21	23	24	25	27	28	30	32	35	44
18	20	21	23	24	25	27	28	30	32	35	46
18	20	21	23	24	25	27	28	30	32	35	46
18	20	21	23	24	25	27	28	30	32	35	46
18	20	22	23	24	25	27	28	30	32	35	49
18	20	22	23	24	25	27	28	30	32	35	49
18	20	22	23	24	25	27	28	30	32	35	49
18	20	22	23	24	25	27	28	30	32	35	52
18	20	22	23	24	25	27	28	30	32	35	53
18	20	22	23	24	25	27	28	30	32	35	66
18	20	22	23	24	25	27	28	30	32	36	
18	20	22	23	24	25	27	28	30	32	36	
18	20	22	23	24	25	27	28	30	32	36	
18	20	22	23	24	25	27	28	30	32	36	
18	20	22	23	24	26	27	28	30	32	36	
18	20	22	23	24	26	27	28	30	32	36	
19	20	22	23	24	26	27	28	30	32	36	

Appendix 10 Average Interply Bitumen in Pitch Built-up Roofing Membranes, lbs per 100 square feet

18	22	23	25	27	28	29	30	31	33	36	42
18	22	24	26	27	28	29	30	31	33	36	42
18	22	24	26	27	28	29	30	31	33	36	42
18	22	24	26	27	28	29	30	31	33	36	43
19	22	24	26	27	28	29	30	31	34	37	43
19	22	24	26	27	28	29	30	32	34	37	
19	22	24	26	27	28	29	30	32	34	37	
19	23	24	26	27	28	29	30	32	34	37	
20	23	24	26	27	28	29	30	32	34	37	
20	23	24	26	27	28	29	30	32	34	38	
21	23	24	26	28	28	29	30	32	34	38	
21	23	25	26	28	28	29	31	33	35	38	
21	23	25	26	28	29	30	31	33	35	38	
21	23	25	26	28	29	30	31	33	35	39	

Appendix 10 Average Bitumen (continued)

21	23	25	26	28	29	30	31	33	35	39
21	23	25	26	28	29	30	31	33	35	39
22	23	25	27	28	29	30	31	33	35	39
22	23	25	27	28	29	30	31	33	36	40
22	23	25	27	28	29	30	31	33	36	41
22	23	25	27	28	29	30	31	33	36	41

Appendix 11 Voids in Interply Asphalt Percent of Sample Area.

0	0	0	0	0	1	1	2	3	3	5	9
0	0	0	0	0	1	1	2	3	3	5	9
0	0	0	0	0	1	1	2	3	3	5	9
0	0	0	0	0	1	1	2	3	3	5	9
0	0	0	0	0	1	1	2	3	3	5	9
0	0	0	0	0	1	1	2	3	3	5	9
0	0	0	0	0	1	1	2	3	3	5	10
0	0	0	0	0	1	1	2	3	3	5	10
0	0	0	0	0	1	1	2	3	3	5	10
0	0	0	0	0	1	1	2	3	3	5	10
0	0	0	0	0	1	1	2	3	3	5	10
0	0	0	0	0	1	1	2	3	3	5	10
0	0	0	0	0	1	1	2	3	3	5	10
0	0	0	0	0	1	1	2	3	3	5	10
0	0	0	0	0	1	1	2	3	3	5	10
0	0	0	0	0	1	1	2	3	3	5	10
0	0	0	0	0	1	1	2	3	3	5	11
0	0	0	0	0	1	1	2	3	3	5	11
0	0	0	0	0	1	1	2	3	3	5	11
0	0	0	0	0	1	1	2	3	3	6	11
0	0	0	0	0	1	1	2	3	3	6	11
0	0	0	0	0	1	1	2	3	3	6	11
0	0	0	0	0	1	1	2	3	3	6	11
0	0	0	0	0	1	1	2	3	3	6	11
0	0	0	0	0	1	1	2	3	3	6	11
0	0	0	0	0	1	1	2	3	4	6	11
0	0	0	0	0	1	1	2	3	4	6	11
0	0	0	0	0	1	1	2	3	4	6	11
0	0	0	0	0	1	1	2	3	4	6	11
0	0	0	0	0	1	1	2	3	4	6	12
0	0	0	0	0	1	1	2	3	4	6	12
0	0	0	0	0	1	1	2	3	4	6	12
0	0	0	0	0	1	1	2	3	4	6	12
0	0	0	0	0	1	1	2	3	4	6	12
0	0	0	0	0	1	1	2	3	4	6	12
0	0	0	0	0	1	1	2	3	4	6	12
0	0	0	0	0	1	1	2	3	4	6	13
0	0	0	0	0	1	1	2	3	4	6	13
0	0	0	0	0	1	1	2	3	4	6	13
0	0	0	0	0	1	1	2	3	4	6	13
0	0	0	0	0	1	1	2	3	4	6	13
0	0	0	0	0	1	1	2	3	4	6	13

Appendix 11 Voids in Asphalt (continued)

0	0	0	0	0	1	1	2	3	4	6	13
0	0	0	0	0	1	1	2	3	4	6	14
0	0	0	0	0	1	1	2	3	4	6	14
0	0	0	0	0	1	1	2	3	4	6	14
0	0	0	0	0	1	1	2	3	4	7	14
0	0	0	0	0	1	1	2	3	4	7	15
0	0	0	0	0	1	1	2	3	4	7	15
0	0	0	0	0	1	1	2	3	4	7	15
0	0	0	0	0	1	1	2	3	4	7	15
0	0	0	0	0	1	1	2	3	4	7	15
0	0	0	0	0	1	1	2	3	4	7	16
0	0	0	0	0	1	2	2	3	4	7	16
0	0	0	0	0	1	2	2	3	4	7	16
0	0	0	0	0	1	2	2	3	4	7	16
0	0	0	0	0	1	2	2	3	4	7	16
0	0	0	0	0	1	2	2	3	4	7	17
0	0	0	0	0	1	2	2	3	4	7	17
0	0	0	0	0	1	2	2	3	4	7	17
0	0	0	0	0	1	2	2	3	4	7	17
0	0	0	0	0	1	2	2	3	4	7	18
0	0	0	0	0	1	2	2	3	4	7	18
0	0	0	0	0	1	2	2	3	4	7	18
0	0	0	0	0	1	2	2	3	4	7	19
0	0	0	0	0	1	2	2	3	4	7	21
0	0	0	0	0	1	2	2	3	4	7	21
0	0	0	0	0	1	2	2	3	4	7	22
0	0	0	0	0	1	2	2	3	4	7	22
0	0	0	0	0	1	2	2	3	4	7	22
0	0	0	0	0	1	2	2	3	4	7	25
0	0	0	0	0	1	2	2	3	4	8	26
0	0	0	0	0	1	2	2	3	4	8	28
0	0	0	0	0	1	2	2	3	4	8	29
0	0	0	0	0	1	2	2	3	4	8	30
0	0	0	0	0	1	2	2	3	4	8	31
0	0	0	0	0	1	2	2	3	4	8	33
0	0	0	0	0	1	2	2	3	4	8	36
0	0	0	0	0	1	2	2	3	5	8	36
0	0	0	0	0	1	2	2	3	5	8	51
0	0	0	0	0	1	2	2	3	5	8	69
0	0	0	0	0	1	2	2	3	5	8	
0	0	0	0	0	1	2	2	3	5	8	
0	0	0	0	1	1	2	2	3	5	8	
0	0	0	0	1	1	2	2	3	5	8	
0	0	0	0	1	1	2	2	3	5	8	
0	0	0	0	1	1	2	2	3	5	8	
0	0	0	0	1	1	2	2	3	5	8	
0	0	0	0	1	1	2	2	3	5	9	
0	0	0	0	1	1	2	2	3	5	9	
0	0	0	0	1	1	2	2	3	5	9	
0	0	0	0	1	1	2	2	3	5	9	
0	0	0	0	1	1	2	2	3	5	9	
0	0	0	0	1	1	2	3	3	5	9	
0	0	0	0	1	1	2	3	3	5	9	

Appendix 12 Voids in Interply Coal Tar Pitch Percent of Sample Area.

0	0	0	0	0	0	1	1	2	3	6	24
0	0	0	0	0	0	1	1	2	3	6	25
0	0	0	0	0	0	1	1	2	3	6	26
0	0	0	0	0	0	1	1	2	3	7	36
0	0	0	0	0	0	1	1	2	3	7	41
0	0	0	0	0	0	1	1	2	3	7	
0	0	0	0	0	0	1	1	2	3	8	
0	0	0	0	0	0	1	1	2	3	8	
0	0	0	0	0	0	1	1	2	3	10	
0	0	0	0	0	0	1	1	2	4	10	
0	0	0	0	0	0	1	1	2	4	11	
0	0	0	0	0	0	1	1	2	4	11	
0	0	0	0	0	0	1	1	3	5	12	
0	0	0	0	0	1	1	2	3	5	23	
0	0	0	0	0	1	1	2	3	5	24	

Rene M. Dupuis

STRAIN ENERGY AND ELONGATION BEHAVIOR OF FIELD SAMPLES TAKEN FROM
POLYESTER REINFORCED BUILT-UP ROOFS

REFERENCE: Dupuis, Rene M., "Strain Energy and Elongation
Behavior of Field Samples Taken From Polyester Reinforced
Built-Up Roofs," Roofing Research and Standards Development:
Second Volume, ASTM STP 1088, Thomas J. Wallace and Walter
J. Rossiter, Eds., American Society for Testing and Materials,
Philadelphia, 1990.

ABSTRACT: A laboratory study was undertaken to determine the
load-strain properties and watertight/strain levels of aged
polyester reinforced built-up roof membranes. Tests were run
at 0°F on field samples of roof membranes ranging in age from
three to five years. One sample of a glass and polyester reinforced membrane was also evaluated. Strain energy values
were determined for all specimens tested. A three point
watertight integrity test was devised, using a prestressing
level of first yield, average ultimate strain, and end point
strain.

KEYWORDS: load-strain, polyester, elongation, watertight
integrity, strain energy density

HOT APPLIED BITUMINOUS MEMBRANES

Bituminous built-up roof membranes have a long history of performance. The use of hot, field applied bitumen with alternating layers
of ply felts dates back well over 100 years.

Up until the late 1970's, built-up roof membranes primarily utilized organic asphalt saturated felts which served as a reinforcing layer
between thin layers of bitumen. By 1980 the use of organic felts de-
clined; fiberglass mats came into wide use. Fiberglass offered superior
tensile strength, but did not improve the elongation behavior which was
nominally 1 to 1.5%. Of interest, however, was the structure of the
fiberglass mat; it is basically open and porous, allowing hot bitumen
to infill the mat structure. Blistering and splitting problems have
decreased dramatically with the use of fiberglass mats.

Dr. Rene M. Dupuis is President of Structural Research, Inc., 3213
Laura Lane, Middleton, WI 53562.

Concurrent with the change from organic felts to fiberglass mats in built-up membranes, the use of modified bitumen membranes grew. In the late 1970's approximately 2% of the market was taken by prefabricated modified bitumen sheets. In 1989, it is estimated that 18% of the entire low slope industrial roofing market will use prefabricated modified bitumen sheets. Of particular interest is the development and use of a variety of spunbond polyester reinforcements in prefabricated modified bitumen sheets. Polyester offers very high elongation potential, generally in excess of 20%; different modified asphalts may affect the elongation while some sheets also incorporate a glass mat.

Since polyester mat has worked so well in prefabricated sheets of modified bitumen, the continued development of bituminous roof membrane systems has now seen polyester used with hot, field applied asphalt. These built-up membranes are typically two plies of polyester mat ranging in weight from 150 to 200 grams per square meter. Asphalts used are typically ASTM D312 Type III mopping asphalts.

Previous research[1][2][3] has documented the load-strain properties of bituminous membrane systems. This study goes beyond previous studies in attempting to find what elongation level causes a membrane to lose watertight integrity, as well as addressing the load-strain properties of aged built-up roof specimens.

TEST PROGRAM

Samples of polyester reinforced built-up roofs were procured. All samples were tested in tension using a Universal Testing Machine at a rate of 0.08 inches per minute at 0°F. The sample descriptions are given below:

A. North Carolina roof age three years. Two-ply spunbond polyester with a fiberglass facer from the polyisocyanurate insulation. A heavy flood coat was aluminum coated; deep alligatoring of the asphalt was present.

B. Wisconsin roof age three years. Two-ply spunbond polyester laid over perlite insulation. The flood coat was aluminum coated and had heavy alligatoring of the asphalt; sample construction was uniform.

C. Wisconsin roof age three years. Two-ply spunbond polyester laid over perlite insulation. The flood coat had heavy alligatoring of the asphalt; sample construction was uniform.

D. Texas roof age three years. Two-ply fiberglass and one-ply spunbond polyester laid over a built-up roof. The flood coat was aluminum coated; sample was extremely thin.

E. Texas roof age five years. Two-ply spunbond polyester laid over wood fiberboard insulation. The flood coat was aluminum coated; sample construction was thin.

Due to the limited size of the field samples, no interply asphalt weights or mat weights could be accurately determined. Relative visual comparisons were made and noted above.

RESULTS OF LOAD-STRAIN TESTS

The tensile, strain, and strain energy data are shown in Table 1. Ultimate tensile strengths ranged from 122 lbs/in to 160 lbs/in for Samples A, B, C, and E. Sample D was of three-ply construction, having two fiberglass plies and one polyester mat.

The ultimate strain levels for Samples A, B, C, and E are all in excess of 30%, while Sample D varies from 1.4 to 2.8% due to the mixture of glass and polyester. The typical load-strain behavior for Samples A, B, C, and E are shown in Figure 1. The highest load shown in Figure 1 is designated as the end point of the test; all load-strain and strain energy values are computed from this point. Sample D, however, has a significantly different load-strain curve as shown in Figure 2. Due to the combination of glass and polyester mats (of undetermined weights), the end point of the load-strain test is taken at a very low point of elongation for the specimen. The glass mat offered a linear resistance to the tensile force until it ruptured; after that, the single ply of polyester continued to elongate, overcoming the localized rupture energy and then continued to increase in resistance until it fractured.

The load-strain properties of the aged polyester reinforced built-up roof membranes are within ranges expected, except for Sample D where a very low tensile strength (83 lbs/in) was seen in the Y direction. A low elongation (1.42%) also accompanied the low tensile strength value.

STRAIN ENERGY VALUES

The strain energy values were computed electronically from the load-strain curves generated, and are listed in Table 1 for each sample. Samples A, B, C, and E all had strain energy density values ranging from a low of 4.19 to a high of 6.70 lb/in/in at 0°F. While there is not specific strain energy density recommended criteria for a 2-ply polyester system, one industry recommendation[3] suggests that a minimum strain energy density be set at 3.0 lb/in/in if tensile strength is below 200 lb/in at 0°F. It should be noted that the 3.0 lb/in/in strain energy criteria recommends that a watertight integrity test be conducted on specimens after partial elongation has been imposed to determine at what elongation loss of watertightness occurs.

WATERTIGHTNESS TEST

Samples A, B, D, and E were selected for water column tests according to the Canadian General Standards Board 37-GP-56M criteria (Standard for: Membrane, Modified, Bituminous, Prefabricated, and Reinforced for Roofing). This test involves the use of a 2" wide membrane specimen, but does not pre-impose a strain on it prior to water column testing. Essentially, the test sample is loaded to a predetermined point of elongation and then mechanically fixed to a backer board while in the elongated or stretched position. A water column is then affixed to the base of the sample for testing.

Table 2 lists the load and strain levels imposed on the BUR specimens prior to water column testing. Since this investigative work is
attempting to find where watertight integrity may be lost, the test
program called for three levels of elongation to be used and as shown
in Figure 3. Water Column 1 is placed on the lowest level of prestress
and Water Column 3 is placed on the highest prestress or near the end
point of the load-strain curve shown in Figures 1 and 2. Each test
specimen is prestressed to a different level.

Due to the presence of glass and polyester in Sample D, a different prestress had to be used. Figure 4 outlines the prestress levels
imposed on Sample D. It should be noted that Sample Y1 had Water
Column 1 placed at a prestressed level of 72 lbs/in which is slightly
less than a 1% strain and Sample Y2 had a prestress level of 92 lbs/in
with an approximate strain in excess of 1% as shown in Table 2.

The results of the varying strain level and resulting watertightness is shown in Table 3.

DISCUSSION OF WATERTIGHTNESS TEST

Procedurally, all specimens prepared for water column testing were
strained at 0°F to the load-strain levels indicated in Table 2. At
that point the test machine was stopped and the specimens were immediately secured with steel bands and metal screws to a 5" X 7" plywood
back board; the specimens were not allowed to relax.

Three of the specimens passed the 20" water column test at low
levels of strain (1-3%). The other 9 specimens exhibited leaks that
would appear to allow water to penetrate the system over time. The
failure of all 9 of these specimens was either immediate or within 15
minutes of the introduction of no more than 5" of water into the column.

Due to the presence of alligatored surface asphalt, extreme diffi-
culty was experienced in maintaining watertight integrity of the water
column seal to the specimen. All of the failed specimens exhibited
moisture penetration past the first ply of polyester, however, it could
not be determined whether the moisture penetrated completely through
the specimens. All failed specimens allowed water at least through
the first ply and out the side (cut edge) of the specimen. Sample E
had many hairline cracks over the surface making the water column test
difficult. The surface coating exhibited severe cracks at 0°F during
the load-strain test (regardless of load level) such that a loose surface coating resulted.

SUMMARY AND RECOMMENDATIONS

The load-strain properties of the aged 2-ply polyester reinforced
built-up membranes show good tensile strength and elongation values.
However, Specimens A, D, and E had highly nonuniform coatings of asphalt along with very thin interply bitumen layers.

Water column tests conducted at various points of elongation on the samples has shown that loss of watertight integrity occurred somewhere between a nominal 2 and 20% elongation. This data is primarily derived from the behavior of Samples A and B.

The loss of watertight integrity of Sample E was due to the numerous hairline cracks present on the surface. This five year old specimen was also thin in composition.

Sample D, which was a combination of glass and polyester, showed a very interesting behavior under the water column test. As shown in Table 3, a prestressed load at less than 1% strain passed a 20" water column, while a small additional load to go slightly beyond the end point as shown in Figure 2 produced a loss of watertight integrity.

Based on the results from the 2-ply specimens, the following observations are in order:

1. Combinations of glass and polyester reinforcing will produce a membrane which has a usable ultimate elongation behavior similar to that of one containing a membrane which has all glass reinforcement. In this test series, Sample D had two plies of glass and one ply of polyester.

2. The ultimate tensile strength failure of the test specimens produced a fine asphalt dust when the specimens ruptured, indicating that the polyester offers a fine reinforcing matrix to the asphalt embedded within it.

3. The usable elongation for the aged 2-ply samples is less than 20%; additional testing would be required to more accurately locate the elongation level commensurate with the loss of watertight integrity.

4. The results of the tests to date indicate that a 2-ply polyester appears to offer adequate load-strain property despite the brittle nature of the mopping asphalt at 0°F.

5. The use of a modified mopping asphalt would greatly enhance the capability of the polyester reinforcement, probably achieving a much higher watertight/strain level.

6. It should be noted that this was a severe test (20" water column) on a 2-ply membrane specimen that was prestressed at 0°F.

7. It is recommended that this technique be adapted for use on the evaluation of modified bitumens in general to more adequately determine what level of usable watertight/strain is present, based on this watertight test approach. Prestress levels should be set at first yield, average ultimate strain, and end point strain.

REFERENCES

[1] Dupuis, R. M. and Lee, J. W., "Coefficient of Expansion Values for Modified Bitumen Roof Membranes," Roofing Research and Standards Development, ASTM STP 959, R. A. Critchell, Ed., American Society for Testing and Materials, Philadelphia, 1987, pp. 92-99.

[2] Lee, J. W. and Dupuis, R. M., "A Performance Evaluation of Modified Bitumen Roof Membranes Using the Strain Energy Approach," Research Report sponsored by The Lutravil Company, Durham, NC, January, 1987.

[3] Rossiter, W. J. and Bentz, D. P., "Strain Energy of Bituminous Built-Up Membranes: A New Concept in Load-Elongation Testing," Applied Technology for Improving Roof Performance - 8th Conference of Roofing Technology, Sponsored by National Bureau of Standards and Published by National Roofing Contractors Association, Gaithersburg, MD, 1987.

TABLE 1. LOAD-STRAIN PROPERTIES AT 0 DEGREES F.

SPECIMEN	TENSILE STRENGTH lb/in (kN/m)	PERCENT STRAIN %	TOTAL STRAIN ENERGY lb/in (kN/m)	STRAIN ENERGY DENSITY lbf/in/in (N/m/m)
A. North Carolina **Coated - 3 Years Old**				
X1	133.86	40.06	41.51	4.96
X2	137.53	38.66	42.02	5.09
X3	133.05	35.71	37.67	4.37
	------	-----	-----	----
AVERAGE	134.81 (23.59)	38.14	40.40 (7.07)	4.81 (21.38)
STANDARD DEVIATION	1.95	1.81	1.94	0.31
Y1	154.32	41.84	50.54	6.63
Y2	163.72	42.88	53.75	6.83
Y3	163.18	40.56	51.61	6.66
	------	-----	-----	----
AVERAGE	160.41 (28.07)	41.76	51.97 (9.09)	6.70 (29.82)
STANDARD DEVIATION	4.31	0.95	1.33	0.09
B. Wisconsin **Coated - 3 Years Old**				
MD1	137.28	28.55	31.80	4.04
MD2	127.99	30.65	32.19	3.96
MD3	148.25	36.51	43.13	5.57
	------	-----	-----	----
AVERAGE	137.84 (24.12)	31.90	35.71 (6.25)	4.52 (20.11)
STANDARD DEVIATION	8.28	3.37	5.25	0.74
XMD1	114.71	28.74	26.73	3.34
XMD2	126.06	33.25	33.54	4.33
XMD3	130.39	37.62	38.58	4.90
	------	-----	-----	----
AVERAGE	123.72 (21.65)	33.20	32.95 (5.77)	4.19 (18.63)
STANDARD DEVIATION	6.61	3.63	4.86	0.64

TABLE 1.   LOAD-STRAIN PROPERTIES AT 0 DEGREES F (CONTINUED)

SPECIMEN	TENSILE STRENGTH lb/in (kN/m)	PERCENT STRAIN %	TOTAL STRAIN ENERGY lb/in (kN/m)	STRAIN ENERGY DENSITY lbf/in/in (N/m/m)
C. Wisconsin Non-Coated - 3 Years Old				
MD1	148.80	35.32	42.91	5.45
MD2	145.07	37.00	41.65	5.37
MD3	152:18	41.79	49.15	6.45
	------	-----	-----	----
AVERAGE	148.68	38.04	44.57	5.76
	(26.02)		(7.80)	(25.60)
STANDARD DEVIATION	2.90	2.74	3.28	0.49
XMD1	118.15	37.23	34.91	4.65
XMD2	127.10	35.60	35.58	4.52
XMD3	121.97	37.64	36.17	4.59
	------	-----	-----	----
AVERAGE	122.41	36.82	35.55	4.59
	(21.42)		(6.22)	(20.41)
STANDARD DEVIATION	3.67	0.88	0.51	0.06
D. Texas Coated 3 - Years Old				
X1	149.99	3.03	2.42	0.31
X2	141.44	2.32	1.90	0.25
X3	174.09	3.03	3.03	0.39
	------	-----	-----	----
AVERAGE	155.17	2.79	2.45	0.31
	(27.16)		(0.43)	(1.40)
STANDARD DEVIATION	13.82	0.33	0.46	0.06
Y1	101.32	1.65	0.88	0.12
Y2	66.96	0.98	0.37	0.05
Y3	81.45	1.63	0.56	0.07
	------	-----	-----	----
AVERAGE	83.24	1.42	0.60	0.08
	(14.57)		(0.11)	(.36)
STANDARD DEVIATION	14.08	0.31	0.21	0.03

TABLE 1. LOAD-STRAIN PROPERTIES AT 0 DEGREES F (CONTINUED)

SPECIMEN	TENSILE STRENGTH lb/in (kN/m)	PERCENT STRAIN %	TOTAL STRAIN ENERGY lb/in (kN/m)	STRAIN ENERGY DENSITY lbf/in/in (N/m/m)
E. Texas				
Coated 5 - Years Old				
X1	146.45	30.20	35.36	4.56
X2	145.34	34.07	40.22	5.27
X3	137.15	28.40	32.10	4.14
	------	-----	-----	----
AVERAGE	142.98	30.89	35.89	4.66
	(25.02)		(6.28)	(20.73)
STANDARD DEVIATION	4.15	2.37	3.34	0.47
Y1	128.36	29.56	30.76	4.10
Y2	143.81	34.22	39.28	5.24
Y3	134.31	34.25	36.73	4.74
	------	-----	-----	----
AVERAGE	135.49	32.68	35.59	4.69
	(23.71)		(6.23)	(20.87)
STANDARD DEVIATION	6.36	2.20	3.57	0.46

NOTE: 1. Test Samples using an"X" or "Y" designation were received without a specified mat direction.

2. Test samples using an "MD" or "XMD" designation were received with a specified mat direction. "MD" denotes machine direction and "XMD" denoted cross machine direction.

TABLE 2 -- LOAD-STRAIN LEVELS IMPOSED ON BUR SPECIMENS
BEFORE WATER COLUMN TESTING

SPECIMEN	LOAD (LB/IN)	APPROXIMATE PERCENT STRAIN
A. North Carolina Coated - 3 Yrs. Old		
X1	84	2
X2	114	21
X3	137	40
B. Wisconsin Coated - 3 Yrs. Old		
XMD1	75	3
XMD2	105	19
XMD3	115	31
D. Texas Coated - 3 Yrs. Old		
Y1	71	1
Y2	92	1
Y3	46	5
E. Texas Coated - 5 Yrs. Old		
Y1	88	4
Y2	115	18
Y3	131	29

TABLE 3 -- STRAIN LEVEL AND RESULTING WATERTIGHTNESS
OF BUR SPECIMENS

SPECIMEN	APPROXIMATE PERCENT STRAIN	WATER COLUMN HEIGHT (INCHES)	WATER COLUMN RESULTS (PASS/FAIL)
A. North Carolina Coated - 3 Yrs. Old			
X1	2	20	PASS
X2	21	20	FAIL
X3	40	6	FAIL
B. Wisconsin Coated - 3 Yrs. Old			
XMD1	3	20	PASS
XMD2	19	20	FAIL
XMD3	31	6	FAIL
D. Texas Coated - 3 Yrs. Old			
Y1	1	20	PASS
Y2	1	20	FAIL
Y3	5	6	FAIL
E. Texas Coated - 5 Yrs. Old			
Y1	4	20	FAIL
Y2	18	20	FAIL
Y3	29	6	FAIL

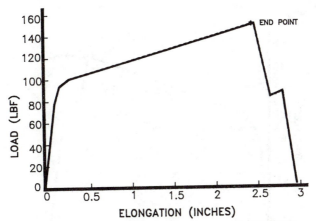

FIG. 1 -- TYPICAL LOAD—STRAIN PLOT OF 2 PLY POLYESTER
SAMPLES A, B, C AND E; SAMPLE GAGE LENGTH
7.5 INCHES.

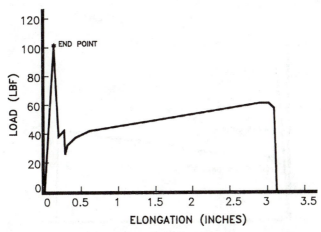

FIG. 2 -- TYPICAL LOAD—STRAIN FOR SAMPLE D, WITH 2 PLY
GLASS AND 1 PLY POLYESTER, SAMPLE GAGE LENGTH
7.5 INCHES.

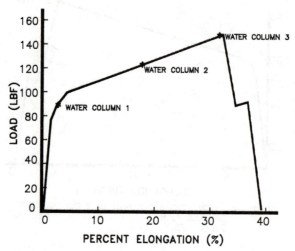

FIG. 3 —— TYPICAL ELONGATION LEVELS IMPOSED ON SAMPLES
A, B AND E.

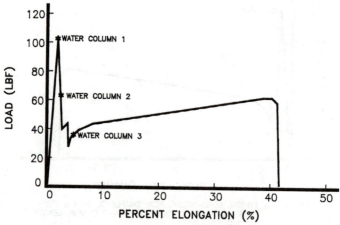

FIG. 4 —— TYPICAL ELONGATION LEVELS IMPOSED ON SAMPLE D.

INDEXES

INDEXES

Author Index

Subject Index